OBLIGATIONS OF SOCIETY

IN THE

XII AND XIII CENTURIES

Oxford University Press, Amen House, London E.C.4

GLASGOW NEW YORK TORONTO MELBOURNE WELLINGTON
BOMBAY CALCUTTA MADRAS KARACHI KUALA LUMPUR
CAPE TOWN IBADAN NAIROBI ACCRA

FIRST EDITION 1946
REPRINTED LITHOGRAPHICALLY IN GREAT BRITAIN
AT THE UNIVERSITY PRESS, OXFORD
FROM SHEETS OF THE FIRST EDITION
1949, 1960

OBLIGATIONS OF SOCIETY
IN THE XII AND XIII CENTURIES

THE FORD LECTURES
DELIVERED IN THE UNIVERSITY OF OXFORD
IN MICHAELMAS TERM 1944

BY

AUSTIN LANE POOLE
FELLOW OF ST. JOHN'S COLLEGE, OXFORD

OXFORD
AT THE CLARENDON PRESS

PRINTED IN GREAT BRITAIN

TO THE MEMORY OF
MY FATHER
REGINALD LANE POOLE
FORD'S LECTURER
1911

PREFACE

THE following lectures are here printed substantially as they were delivered in Michaelmas Term, 1944. There has been some slight rearrangement of the material and, in view of the fact that I am engaged on a book covering wider aspects of the period, some paragraphs which seem to belong more appropriately to the forthcoming volume have been omitted.

Unfortunately Mr. C. T. Flower's elaborate and important introduction to the *Curia Regis Rolls*, published by the Selden Society as the volume for 1944, was only issued to subscribers in July 1945, after these lectures had gone to press. I have therefore not had the opportunity of profiting by this valuable commentary. Mr. Sidney Painter's *Studies in the History of the English Feudal Barony* (The Johns Hopkins Press, Baltimore, 1943), which deals with some questions here discussed, also reached me too late to avail myself of it.

I have received much help and encouragement from many friends to whom I wish to express my deep sense of gratitude. Particularly I should like to thank Professor and Mrs. Stenton who, not only placed with rare generosity their unrivalled knowledge of the period at my service, but were also kind enough to read the lectures in manuscript.

A. L. P.

October 1945

CONTENTS

I

THE CLASSIFICATION OF SOCIETY

IN the interval of twenty years which separates this war from the last the Public Record Office has made available to students of history the Plea Rolls, the Rolls of the King's Court, for the reigns of Richard I and John in seven magnificently edited volumes.[1] Side by side with this official publication the Pipe Roll Society, thanks to the unsparing work of Mrs. Stenton, has carried forward the printing of the Exchequer accounts, which have now reached to the eighth year of King John, the year ending at Michaelmas 1206.[2] In the same period was printed the *Book of Fees*. This was kept at the Exchequer as a book of reference from which the clerks could ascertain the details they required concerning tenures and services. It has indeed been long in print under the title *Testa de Nevill*, but the edition of the Record Commission published in 1807 was as likely to mislead as to assist scholars. To quote Sir Henry Maxwell Lyte, the late Deputy Keeper of the Public Records:

'In the edition of 1807, sections different in character and in date are often printed consecutively, without even a note to suggest that there should be an interval between them. Conversely, blank spaces, emphasized by lines across the text, have been introduced at haphazard in sections which should proceed continuously. The most charitable theory is that the editor left the printers to deal with the transcript as they pleased. Anyhow, the result is chaotic.'[3]

Such was the opinion of the man under whose direction appeared between the years 1920 and 1931 the splendid new edition. If I remind you that the monumental index of persons and places runs to 699 large octavo pages printed in double columns, the comprehensiveness of this fundamental source of feudal history will be apparent.

This wealth of new material of unrivalled importance for the social history of the twelfth and thirteenth centuries must be

---

[1] *Curia Regis Rolls of the Reigns of Richard I and John preserved in the Public Record Office*, cited as *C.R.R.* in the following pages.
[2] Cited as *P.R.* with the Exchequer year of the reign.
[3] *Book of Fees*, i, pp. vi–vii.

my excuse for choosing a subject for discussion which has been so thoroughly and so finely treated by the best scholars of the last generation, by Maitland, Round, and Vinogradoff.

In the period to which these records relate law was still in the making; it had not yet reached anything near the precision of Bracton's time. The lawyers are laying down the law with few rules to guide them; for, great as the Norman and Angevin kings were as administrators, they were not great legislators. They brought no code of law with them, and they did not, like their Anglo-Saxon predecessors, Alfred, Edgar, or Cnut, issue any. The judges administering the law have to act not on formal enactments but on instructions, sometimes, no doubt, verbal instructions, given them by the king or his advisers. The decisions of courts may have been recorded on rolls from the earliest years of Henry II's reign, but none of these early rolls have survived. The officials at the Exchequer were only interested in the amercements imposed by the courts, and when these had been entered on the Pipe Roll, the rolls of the justices had served their purpose. The introduction of a more methodical system for the preservation of plea rolls may probably be attributed to the administrative genius of Hubert Walter. The first extant roll belongs to the year 1194, the year in which Hubert became justiciar. But there is little to suggest that they were as yet used (at any rate extensively) as a record of legal precedents.[1] In these circumstances there is much room for difference in practice, in procedure, in decisions.

When in the first quarter of the twelfth century a Norman lawyer tried to write down what he thought was the law in his time, he wrote about class distinctions thus: 'There is a distinction of persons in condition, in sex, according to profession and order, and according to the law which should be observed, which things must be weighed by the judges in all matters.'[2] At the end of the century the lawyers were trying to evolve a much more simple classification of society. This was according to the mode of tenure of land and the obligations inherent in that mode of tenure. A man must be a knight owing military service, or a sergeant owing some specific service; he must be

---

[1] See Mrs. Stenton's introduction to *The Earliest Lincolnshire Assize Rolls* (Lincoln Record Society, vol. xxii), p. xxii.

[2] *Leges Henrici Primi*, 9. 8; Liebermann, *Die Gesetze der Angel-Sachsen*, i. 555.

a socager paying rent or some render in kind; or he must be unfree and do servile work; or, grandest of all, he or some religious corporation may hold by simply praying. The lawyers maintain that a man must fall into one or other of these groups, and that his status can be easily determined by one or two simple tests. A knight can be distinguished from a sergeant or a socager by the fact that he was liable to pay scutage, the money commutation for military service in the field. If, says Bracton, he pays but a halfpenny in scutage, his service is military.[1] The status of a villein can be determined by whether or no he performs uncertain agricultural services and pays a small sum called merchet when he gives his daughters in marriage. The division of tenures into frankalmoign, knight service, sergeanty, socage, and villeinage Maitland regards as already in the thirteenth century becoming the classical legal scheme.[2]

The records of the late twelfth and early thirteenth centuries suggest that an arrangement of society on such a basis is too orderly, too logical. These fine distinctions between classes do not altogether correspond with the facts. They are an over-simplification: the artificial creation of lawyers. As long ago as the year 1900 the late Sir Paul Vinogradoff, who was both lawyer and historian, wrote in a paper contributed to the *Economic Journal*:

'It would have been pleasant and clear indeed, if the whole of English society could have been arranged under the headings of villeins holding by rural work, socagers holding by rent, knights and sergeants holding by military service, clergy holding by ecclesiastical obligations. The reality of things did not quite admit of such simplicity.'[3]

The reality of things is certainly not simple; it is full of complexities. The classical scheme, as Maitland calls it, even before it has reached positive definition, has already become a little out of date. This is one of the difficulties with which one is faced in treating of the organization of feudal society. Before it has reached perfection as a system the seeds of decay have been sown and have germinated. While in one aspect it is still developing, in another it is rapidly falling into obsolescence.

Already in this period a society based on tenures and services

---

[1] f. 37.   [2] Pollock and Maitland, *Hist. of Eng. Law*, i. 239–40.
[3] *Collected Papers*, i. 127.

is beginning to pass into a society based on money, on rents and taxes. The services on which the various tenures depend are being commuted for money payments. An ever-increasing number of knights acquit their service by the payment of scutage or fine; the special duties of sergeants are 'arrented' or exchanged for rents. Even the labour services of villeins are commuted into money payments. When in the thirteenth century the king tried to compel men possessed of a certain amount of landed property to become knights by writs of distraint, it was not because he was anxious to increase his standing army, nor yet the class of county aristocrats. He doubtless had in mind the pressing need of increasing the number of men available for serving on grand assize juries and performing the other administrative functions which were restricted to the knightly class. But the chief object was financial; to gain more money from those fruitful sources of royal revenue, scutages, reliefs, wardships, and marriages, incidental to military tenures. So early does a money economy intrude itself upon the feudal organization of society.

Scutage is regarded as the distinguishing feature of knight service. It is of course essentially a military payment; it is shield money, a composition for military service. But as a test it is far from satisfactory. The under-tenants, who had no military obligations whatever, were called upon to contribute to scutage; we even hear of villeins contributing to scutage. In 1230 the king directed the sheriff of Lincolnshire to assist one of his tenants-in-chief in compelling 'his knights and free tenants' to render their scutage.[1] In a record of an inquest of 1242-3 it is stated that the abbot of Osney holds half a hide of land in frankalmoign at Hensington near Woodstock, and the clerk adds, as though with surprise, 'and he does not give scutage'.[2]

Now to understand how it is that persons, or rather we should say lands, are burdened with services which are quite foreign to the mode of tenure, we must bear in mind some obvious facts about subinfeudation. By the end of the twelfth century little land was, feudally speaking, held by a single tenant. Between the king, the owner of all the land in England, and

[1] *Memoranda Roll 14 Hen. III* (Pipe Roll Society, N.S. xi), p. 91.
[2] *Book of Fees*, p. 831.

the man who actually cultivates it there are generally several mesne tenants. Often these tenants are holding by different forms of service. Subinfeudation has played havoc with any neat arrangement. Anyone who reads even a few pages of that remarkable work by the late William Farrer, *Honours and Knights' Fees*, will realize the bewildering extent of these ramifications. Let me take a simple case, simple because there is only one service—knight service—involved. In Redmarley, a hamlet in Worcestershire, there was half a knight's fee 'which Adam of Redmarley holds of John son of Geoffrey, John of William of Beauchamp, William of Beauchamp of William de la Mare, William de la Mare of the earl of Gloucester, the earl of the lord king'.[1] Here we have no less than five sub-tenants below the king. When different forms of tenure are involved the case becomes more complex. Simon of Chauncy, a member of a prominent family in the north-east of England, held in 1166 five knights' fees in Lincolnshire of the king. Before 1168 he had granted half one of these fees, which lay at Willoughton, to the Knights Templars in free alms; and the little estate became a preceptory, an administrative centre of the order. The land was carved up into small holdings, normally consisting of a bovate and a toft, and leased to peasants who rendered 5s. and four hens and four days' work a year.[2] Here the king, the Chauncy family, the Templars, and the peasants all have an interest in the same piece of land; the selfsame acres are held by knight service, by frankalmoign, and by rent.

It would be wrong, however, to suppose that the burden of these various services necessarily fell on those who were primarily responsible for them. A landowner who granted away some of his land made a contract with his tenant for the performance of some service—it might be a portion of knight service, it might be a spiritual service, or a rent, or something of the kind. This is technically known as the intrinsic service, for it is inherent in the bargain. But he may also pass on to his new tenant the service for which he is himself responsible to his overlord. This is outside the contract, and is known as forinsec service. If a particular tenement is burdened with

[1] Ibid., p. 961.
[2] *Records of the Templars in England in the Twelfth Century*, ed. B. A. Lees, pp. 100–1, Introduction, pp. cxci–cxciv.

more than one service, the difficulty of classifying society on the basis of tenure and service is considerably increased. Let me illustrate this from that form of tenure which ranks first in order of superiority, and which by its very nature should be easiest to detect—frankalmoign.

In an age of violence tempered with pious superstition men were accustomed to endow churches in the hope of future salvation. Many thousands of acres passed in this way during the twelfth and thirteenth centuries into the hands of parish churches and religious houses. The grants are usually expressed as gifts 'in free alms' (*in elemosinam*) or more emphatically 'in free, pure, and perpetual alms'. And the donor goes on to say that he does this for the safety of his soul or the safety of the souls of his relatives and ancestors. But these categorical statements are apt to be misleading, for he will not infrequently add, as though somewhat as an afterthought, that he expects something in return besides merely praying. When, for instance, about the turn of the century, a certain Ralph, son of William of Adewich, granted to the monks of Kirkstall half a carucate of land in Bessacar in Yorkshire, he did so in the following terms:

'Know all, both present and future, that I Ralph, son of William of Adewich, have given and conceded, and by this my present charter have confirmed to God and St. Mary and to the monks of Kirkstall for ever, for the love of God and the safety of my soul, and of the souls of my heirs and of all my ancestors, half a carucate of land in Bessacar, to be held of me and my heirs in pure and perpetual alms, free and quit of all earthly service and secular exaction, with all appurtenances and liberties and easements of the said vill of Bessacar, in wood and plain, in meadow and pasture, in moor and marsh, in highways and byeways, within vill and without, and in all places without any reservation; by rendering to me and my heirs annually for all service, eight shillings, four at Pentecost and four at the feast of St. Martin. And be it known that the monks shall do the forinsec service which belongs to half a carucate of land where twelve carucates make the fee of one knight.'[1]

Such a grant is anything but 'free and quit of all earthly service and secular exaction'. It has both intrinsic and forinsec services of a secular kind attached to it, over and above the spiritual

[1] W. Farrer, *Early Yorkshire Charters*, ii. 165.

service of prayer. If it were not for the words in the charter 'in pure and perpetual alms' and the reference to the souls of the king and his ancestors, there would be little to distinguish it from a tenure in socage. Conversely, a tenure which is burdened only with the work of prayer—the characteristic of frankalmoign—may in fact be classed in a different category. A tenant held a virgate of land at Pusey in Berkshire in chief of the king by the service of saying the Lord's Prayer five times daily for the souls of the king's ancestors.[1] The family acquired the surname Paternoster for their pious work. This was surely a spiritual service; yet it is reckoned not as frankalmoign, but as a sergeanty. The frequent recourse made to the assize *utrum*, a legal process to determine whether a particular tenement belonged to free alms or to lay fee, is evidence of the uncertainty which often prevailed.

From the point of view of classifying society on the basis of tenure, the tenure of sergeanty gives rise to the greatest confusion. Generally speaking, it implies some special service, a service connected with the king's household or his government or his sport or his army. Sergeanty is very promiscuous; it impinges on all other classes. Sometimes, as in the case of the tenant at Pusey, it was barely distinguishable from frankalmoign. More often it approximated to military tenure or socage. Occasionally a mere payment of a small rent could be deemed sergeanty. A remarkable instance of this confusion is afforded by a group of tenants of the bishop of Lincoln in the hundred of Dorchester-on-Thames. They were in Edward I's time performing agricultural services of a normal kind. But the local jurors asserted that their ancestors used to be freemen like sokemen (*quasi sokemanni*), and rendered service to the king for forty days in time of war at their own cost armed with doublet, lance, and steel cap.[2] Now this was the most characteristic service demanded from military sergeants, and yet these men were designated socagers.

These things may at first sight seem to be of mere academic interest. What after all did it matter whether a man was classed as a knight or a sergeant, a socager or a villein? When compared with tenure, Maitland has said, status is unimportant.

[1] E. G. Kimball, *Serjeanty Tenure in Medieval England*, p. 118, n. 71.
[2] *Rotuli Hundredorum*, ii. 748–9; Pollock and Maitland, *Hist. of Eng. Law*, i. 295.

This, no doubt, is in a sense true. But the two things are almost inextricably bound together. The question of status was not infrequently an issue in the courts of law. For a variety of reasons it was a matter of very practical importance both to the lord and the tenant. On it depended the burdens to which a man was subjected. The relief, for example, of a knight was assessed on a different basis to that of a baron or a sergeant; the marriages of widows and children of knights and sergeants were controlled by their lords, while those of tenants in socage were not. Different rules on matters like wardship or the disposition of property obtained in the different classes. The widow of a freeman was entitled to dower (usually a third of the property);[1] a widow of a villein was not, though she could inherit the whole. Different rules again governed the age at which children reached man's estate and were thereby able to dispose of their land. Military tenants came of age at twenty-one, tenants in socage at fourteen, and townsmen often according to the special customs of the borough.[2] Jury service was restricted to freemen, and knights were burdened with other functions in local administration from which the ordinary freemen were exempted.[3]

Both tenure and status were obviously important. But it was not an easy task for the administration to keep track of all the changes in this fluid society. When in 1213 Guy de Chancels accounted for the scutage of the honour of Gloucester, he had to admit that twenty fees 'cannot be found'.[4] It was the duty of the Remembrancers to keep notes about debts to the Crown which had not been paid or about which some difficulty had

---

[1] It was often reckoned with great precision. In a plea of dower, for example, a certain Matilda claimed 'terciam partem iii virgatarum terre et dimidie cum pertinentiis in Bifeld' et terciam partem ii virgatarum terre et dimidie et quarte partis i virgate terre cum pertinentiis in Creulton' et terciam partem tercie partis i molendini cum pertinentiis in eadem villa'. *C.R.R.* vii. 120.

[2] Cf. a Coventry case of 1221 when the jurors said 'the custom in a borough is that he who can count pennies and measure cloth is of age': *Rolls of the Justices in Eyre for Gloucestershire, Warwickshire and Staffordshire, 1221–1222* (Selden Society, vol. lix, no. 574). The number of pennies he was expected to be able to count was normally twelve. See the customs of Hereford (temp. Edward III) printed by W. Wotton, *Leges Wallicae*, p. 517.

[3] A woman in 1219 defended her husband's freedom on the ground that 'ut liber homo fuit in recognicionibus et assisis coram Justiciariis': *Rolls of the Justices in Eyre for Lincolnshire 1218–19 and Worcestershire 1221* (Selden Society, vol. liii, no. 802).

[4] Madox, *Exchequer* (ed. 1711), p. 445 (o).

occurred. The earliest Memoranda Roll which has survived
relates to the year 1199, the first year of King John, and it
illustrates very vividly the difficulties in which the officials were
constantly placed. So under Oxfordshire we read 'Henry son
of Geoffrey owes half a mark of his fine for warranty of his
charter. The sheriff says that he does not know who he is or
where he lives', and a note is added 'let it be inquired who he
is'.[1] Or again, 'William of Bugedenn, steward of the bishop of
Lincoln, promised to make payment of the bishop's scutage,
but the sheriff says that he does not know from how many
fees.'[2] The justices, when they made their periodical circuits
round the country, were specially instructed to investigate how
the crown tenants held their lands. The question is put to
juries of neighbours; often the answer is *nescitur per quod servitium*.

A further source of confusion was the paucity of Christian
names in common use. Some men were blessed with distinctive
names such as Pentecost or Hippolytus; but the vast majority
of the inhabitants of England in the twelfth century shared
among them some two or three dozen names such as John,
Henry, Hugh, Richard, Geoffrey, Robert, or William.[3] Surnames were coming into use to give a distinctive mark to an
individual. These were descriptive of place, profession, or personal characteristic. Ricardus de Dikelesdon (Dixton), Ricardus
de Aneford (Andoversford), Ricardus de Halling (Hawling),
neighbouring hamlets in Gloucestershire; Adam Faber (smith),
Adam Pictor (painter), Adam Horsdriuere (coachman);[4] Herbertus cum barba, Herbertus Inimicus Dei; Moses le riche,
Moses cum naso. But far the commonest way of designating
a particular person was to give the father's name: Gaufridus
filius Petri (or Geoffrey Fitz Peter as we usually speak of
King John's famous justiciar). But with so few names in general
use, this was often inadequate, and the grandfather's and even
the great-grandfather's name might have to be added. So in
a record of the year 1199[5] we find an entry that 'Rogerus filius

---

[1] Pipe Roll Society, N.S. xxi, p. 55.   [2] Ibid., p. 56.
[3] A greater variety is found in the northern and eastern districts where Scan-
dinavian influences prevailed.
[4] Owing to the lack of the definite article in Latin, it is impossible to say
precisely when these professional names became mere surnames, when Adam the
Smith became just Adam Smith.
[5] *Memoranda Roll 1 Jo.*, pp. 38–9.

Nicolai filii Roberti filii Harding' owes £125. 6s. 8d. for his relief. The next entry refers to his uncle as 'Mauricius filius Roberti filii Harding'.

The trouble that might arise from this system of nomenclature may be illustrated from a series of proceedings recorded on the Memoranda Roll of 1230. They begin with a writ to the sheriff of Essex ordering him to hold an inquiry

'whether a loan of 60s. which is demanded by summons of our Exchequer in your county from Richard son of William in the time of lord John the king our father, was made to Richard son of William of Stapelford or to Richard father of Robert of Tilbury; and if the same Richard, the father of Robert himself, was called Richard son of William or Richard son of Robert'.[1]

The next entry relating to the matter states that:

'The attorney of Richard son of William against Robert son of Richard of Tilbury is Ralph de Hawilla or Robert son of the same Ralph for the plea that is between them concerning loans (60s.) made in the time of King John to Richard son of William on account of the identity of name.'[2]

The result of the inquiry is given in a mandate to the sheriff:

'that he may allow Richard son of William of Stapleford to be in peace concerning the loan of 60s. made to Richard son of [Richard cancelled] William in the time of King John, because it is agreed by the barons by inquisition made by order of the king that that loan was not made to the said Richard, but to Richard father of Robert of Tilbury.'[3]

The final stage is reached when the debt is transferred on the Pipe Roll from Richard son of William to the account of Robert son of Richard of Tilbury.[4]

Mistakes over names might sometimes be explained as *per errorem clerici*.[5] But the trouble went deeper than mere clerical errors. The officials got their information about these people from the neighbourhood in which they lived, where they certainly were not known by a cumbersome string of family names. When surnames became fully established some aristocratic families might call themselves Fitz Thomas or Fitz Richard or

---

[1] *Memoranda Roll 14 Hen. III*, p. 78.
[2] Ibid., pp. 88–9.
[3] Ibid., p. 69.
[4] *P.R. 14 Hen. III*, p. 159.
[5] *Memoranda Roll 14 Hen. III*, p. 94.

Fitz Henry, but the common folk became Thomson, Dickson, and Harrison, which suggests that diminutives were even commoner in the Middle Ages than they are to-day.[1] The history of surnames will tell us that the fathers of men who are described as *filius Thomae*, *filius Ricardi*, and *filius Henrici* were generally known to their friends and neighbours as Tom, Dick, and Harry. The language of the records differed from the language of the people. The name in a record and the name by which a person was commonly known were so different that it is remarkable not how little but how much is known and can be traced about the tenure, service, condition, and obligations of the less prominent members of society.

[1] See, for a discussion on the origin of surnames, the introduction by G. J. Turner to his *Calendar of the Feet of Fines relating to the County of Huntingdon*, pp. xvi. ff.

II

THE PEASANTS

THE villeins, *villani*, villagers, form the largest group in the structure of society in medieval England. If we associate with them their congeners, the bordars and cottars, that is, the smaller peasantry, they represent approximately 70 per cent. of the population recorded in Domesday Book. These peasants were the descendants of the free but dependent ceorls of Anglo-Saxon times. The class, however, has swollen in the twenty years that followed the Norman Conquest by the inclusion of recruits both from above and from below. The explicit statement in the Domesday of Essex[1] that there was in 1066 'a certain free man with half a hide who now has become one of the villeins' unquestionably indicates a process which was happening all over the country on a large scale. There was, for example, a substantial class of men in Sussex described as *liberi homines* in the time of King Edward. Twenty years later these have disappeared; they have apparently been absorbed into the ranks of the villeins. On the other hand, if some were degraded, others were exalted. When William Leuric acquired the little Gloucestershire village of Hailes from its Saxon owner Osgot, there were twelve slaves (*servi*) on the establishment;[2] these the new lord made free. As there is in England no separate class of freedmen, we must assume that they also joined the villeins. Indeed, from this source the class continued to grow after 1086. The slaves, male and female (*servi* and *ancillae*), form a not insignificant group in the Domesday record. There were some 26,000 of them. In the earliest surveys of the twelfth century there is hardly a trace of such a class. After the Conquest, economic and humanitarian motives, both the convenience of masters and the preaching of the Church, worked for the abolition of slavery in the proper sense of the term. For the villeins are not slaves. Though we are accustomed to regard them as personally unfree, as they certainly were in the thirteenth century,

[1] *D.B.* ii. 1 *b*.
[2] Ibid. i. 167 c. Cf. also numerous Essex entries, for example Great Henny 'then (1066) and afterwards two *servi*; now none': ibid. ii. 74.

it is at least an open question whether the Domesday commissioners so regarded them. Yet their tendency to sink rather than to rise in the social scale becomes soon apparent. The emancipation of a large number of slaves must have inevitably affected the social position of the class into which these were thrust. It tended to drag down the villeins who now become the lowest class. The abolition or the virtual abolition of slavery was in fact the first great impetus which drove the peasantry on the downward path to serfdom.

But a stronger influence affecting the condition of the peasantry was the development of the common law in the twelfth century, the great heritage bequeathed to us by Henry II. Before this time the question of status did not arise. The distinction between free and unfree had been one between slaves and the rest of society, not between the villeins and the rest of society. The slaves as a class have ceased to exist, and their place on the bottom rung of the social ladder has been taken by the villeins. The king did not wish, perhaps did not dare, to interfere with the private jurisdiction which a lord had over his villeins, who were in a sense his chattels.[1] He would only deal with the wrongs of freemen. So the sharp distinction between freeman and villein begins to emerge in the twelfth century with the growth of royal writs and possessory assizes. It became necessary to decide who could and who could not get the protection of his rights and property in the king's court. But the line is not easy to draw. The lawyers are striving in the face of great difficulties to reduce the whole population into the simple classification, free or serf, *aut liberi aut servi*. Ultimately they were successful in degrading most of the peasants into a condition of serfdom.[2] They did so by the

[1] Cf. the comment in a dispute between the men of Crondall, Hurstbourne Priors, and Whitchurch, and the Prior and Convent of St. Swithun's, Winchester: 'Et Dominus Rex non vult se de eis intromittere', Bracton's *Note Book*, pl. 1237; Vinogradoff, *Villeinage*, p. 46.

[2] This generalization applies particularly to the midland and southern districts. In East Anglia and the north-east of England, where Danish influences prevailed, free sokemen survived in large numbers. In Lincolnshire in fourteen out of the thirty-three wapentakes more than 50 per cent. of the peasantry were free sokemen, and in two the proportion was over 70 per cent. But as we move westward the proportion decreases rapidly. Of the eighteen wapentakes of Leicestershire, Nottinghamshire, and Derbyshire (all counties of the northern Danelaw) in only one was there an excess of sokemen over the villein population, and in two there were no sokemen at all. See F. M. Stenton, 'The Free Peasantry of the

application of certain artificial tests, two of which became in the thirteenth century the standard tests—uncertainty of services and the payment of merchet. But the proof of villeinage was a very intricate matter, and, as the legal records show, one on which there was often much room for doubt.

What, then, was a villein in the eyes of the law? A Northamptonshire peasant in 1198 defends his freedom by declaring that he 'non est rusticus nec servus nec villanus nec nâtus in villenagio', nor has he ever done servile works or customary services.[1] These are the legal terms to denote villeinage. The fine distinctions of the dependent class which we meet with in Domesday Book, bordars and cottars, reappear in manorial records, but they are unknown to the law of the twelfth century. The peasant is a villein, a member of the village community; he is a native, a villein by birth; he is a simple rustic. Sometimes he might be termed a *consuetudinarius*, a man who performs customary services; sometimes he is called *servus*, which must be translated serf not slave, a man who does servile work. Yet many aspects of medieval serfdom are very like slavery. I need not here enumerate the rights which a lord had over his villein. In theory, at least, he could do what he pleased with him except kill him or mutilate him. He could, for example, sell him to another. There are occasional records of such transactions. In 1205 the abbot of Waltham bought a villein for £2; a woman was sold in Norfolk in 1207 for 4s.; a little earlier, in the reign of Henry II, a rustic with his whole family was sold in Lincolnshire for 22s.[2] An interesting charter, which probably belongs to the first part of the thirteenth century, actually grants a villein to the abbey of Sulby in Northamptonshire in frankalmoign:

'Know, both present and future (the charter runs) that I, Hugh of Ringston son of Adam of Ringston, have given and granted and by this my charter have confirmed to God and St. Mary and to the abbot and convent of Sulby, for the safety of my soul, and of my ancestors and successors, in pure and perpetual alms, Robert, the son of Juliana of Walton, with all his brood (*sequela*) and all his chattels, &c.'[3]

Northern Danelaw', *Bulletin de la Société Royale des Lettres de Lund*, 1925–6, especially the tables setting out the percentage of sokemen to villeins and bordars, pp. 77–9.

[1] *C.R.R.* i. 67.

[2] Ibid. iv. 37; v. 94; Stenton, *Danelaw Charters*, no. 273 and Introd., p. lxxxi.

[3] Madox, *Formulare*, p. 417. Hugh of Ringston witnesses charters *circa* 1218 and in 1227; *Registrum Antiquissimum* (Linc. Rec. Soc.), i. 173, ii. 85. He was

Hugh of Ringston is here, so to speak, putting Robert and his *sequela* into the alms-dish; he is treating him like any other piece of property that he might sell or give away. Sometimes it might appear as though the villeins had some say in the matter. The transfer from one owner to another of two virgates of land and of two villeins in Garsington in Oxfordshire was the subject of a final concord in 1253. The property was to be held by Hugh of Garsington by the annual render of ½ lb. of cummin.[1] The final concord was of course a particularly binding form of conveyance, and every precaution was taken. In this case the villeins were present and acknowledged themselves to be villeins. This fact, however, does not imply that they were in any way consenting parties to the contract. Their presence and acknowledgement was simply an additional security.

Nevertheless, in practice if a villein carries out his appointed tasks and does what custom demands of him, then the custom of the manor will normally protect him. The custom of the manor is, in the villein's world, at least nine points of the law. If he does not carry out his customary duties, it seems reasonable enough that he should be deprived of the lands which he holds in return for performing such duties. A certain Edith, for example, who came into possession of a villein tenement by inheritance from her husband, was unwilling to do the customary services. She was accordingly disseized by the lord. But she has no just cause of complaint, and the king's court upheld the lord's action.[2]

Labour services are the essence of villeinage. The villein must work for his lord for so many days in the week in addition to boon works, special tasks, *precaria*, at certain seasons of the year. The technical way of expressing this was that he holds 'by fork and flail', *ad furcam et flagellum*. The lawyers, as I have said, lay stress on the uncertainty of these agricultural services. In fact they were generally known and often minutely specified in writing. The villein does what is customary; he is a *consuetudinarius*. This, surely, is the very antithesis to one who always lives in uncertainty as to what he has to do the next

presumably the brother of Elias son of Adam who held land at Ringston in 1212 (*Book of Fees*, p. 180; cf. also p. 1027) and who witnessed a charter of the late twelfth century (*Danelaw Charters*, no. 411).

[1] *Feet of Fines for Oxfordshire*, ed. Salter (Oxfordshire Record Society), p. 163.
[2] *C.R.R.* vi. 355.

day. Let me take by way of illustration the works which the
peasants on the estate of the Templars at Guiting in Gloucestershire were required to do for their virgate of land in 1185:

'Each virgate of land which owes services must work with one
man for two days in each week from the feast of St. Martin (11
November) till the time of haymaking, and then they will mow for
four days a week as long as there are meadows to be mown and hay
to be carried. If they shall have been mown and the hay carried
before the feast of St. Peter ad Vincula (1 August), they shall return
to working two days a week until St. Peter ad Vincula, and after the
feast of St. Peter for four days a week, unless the corn crops are so
forward that they can reap them, and if they can reap them, then on
Monday they must work with two men and on Tuesday with one
man and on Wednesday with two men and on Thursday with one
man until the corn is carried, and when the corn has been carried,
four days a week until the feast of St. Martin. Further, each virgate
which renders work must plough as a boon work (*de bene*) an acre
and three-quarters, and thrash the seed corn, and sow the land and
harrow it for the winter sowing; and, if the master wishes it, carry
loads to Gloucester or wherever he wills. Each team must also
plough two acres of pasture. All the labourers must also make a
load of malt against Christmas, and similarly against Easter, and
for drying the malt, they must get one load of wood; the said
labourers must also move the sheep-fold twice in the year, and they
must spend two days in washing and shearing the sheep.'[1]

This is a fairly comprehensive list of services, and covers
many of the operations incidental to a farming year. In the
thirteenth century these lists of services are recited in even
greater detail. Of course agricultural work was more than any
other affected by weather conditions. The ground on a certain
day might be unfit for ploughing, and the villein might be put
on to hedge or ditch instead. But it is obvious that he knew
pretty well in the evening what he was likely to have to do the
next morning. Uncertainty of services, therefore, on which the
lawyers insist, amounts to little more than that the villeins, like
most other men, were subject to orders. They must do as they
were told. Bracton's test of villeinage is, to say the least, an
unsatisfactory and inconclusive test.[2] And this fact explains
why on the Plea Rolls we find both parties having recourse to

---

[1] *Records of the Templars in England in the Twelfth Century*, ed. Beatrice A. Lees, p. 50.
[2] Cf. Vinogradoff, *Growth of the Manor*, 349 f.

all manner of other tests to prove this vital question *utrum sit villanus?*

The question of determining villein status is rendered more complicated by the existence of two classes of villeins who were in an exceptional position. I refer to those on 'ancient demesne' of the Crown, and to those who, though personally free, held their lands in villeinage. Let us consider the first of these classes. If by 'ancient demesne' had been meant just the lands belonging to the Crown, the matter would be simple enough. Unhappily it does not. By the term 'ancient demesne' is meant those lands with which William I at the time of the Conquest had endowed the English Crown; the lands which had since that time fallen to the Crown by way of escheat do not form part of the ancient demesne. On the other hand, lands which were originally ancient demesne do not cease to be so regarded if the king has granted them away to others. The rule is that, if lands are recorded in Domesday Book as lands held by the king, then they are and always will be ancient demesne, whoever the present owner may be.

The villein tenants on these estates, villein sokemen, as Bracton calls them, though they are not so conveniently labelled in the records, enjoyed many immunities and privileges denied to the ordinary villeins. They were relieved of the public burdens incumbent on the regular tenant in villeinage; they could leave their tenements when they wished; and they were protected by special writs provided to meet their case both against ejectment from their holdings and against any increase in their services.[1] The privileged position of this class may be illustrated from the manor of Ewell in Surrey. This manor had belonged to the Crown as ancient demesne. Henry II granted it to the canons of Merton priory. In 1202 a certain Simon disputed the possession of some of the land within the manor by an assize mort d'ancestor. The evidence submitted by the prior is significant. He declared that all the tenants of the manor had been villeins of the king and now were villeins of the prior; and that it was never customary for an assize or sworn inquest to be held in the presence of the justices concerning the estates of the manor; but that a sworn inquest

[1] The little writ of right close and the writ *Monstraverunt*. See Pollock and Maitland, *Hist. of Eng. Law*, i. 385 ff., and below, p. 26.

should be held among the villeins on the manor itself according
to the custom of the manors of the king. Judgement was therefore given against Simon, and the assize was not allowed to
proceed.[1] That this was, or at least became, the recognized
law is shown by a number of cases heard on the eyres of the
earliest years of Henry III's reign, several of which attracted
the attention of Bracton. Thus the tenants of King's Norton
in Worcestershire in 1221 successfully pleaded in an assize mort
d'ancestor that the land was 'the king's demesne manor where
no writ of assize lies because it is the king's villeinage'.[2] The
curious position of the villeins on ancient demesne can perhaps
be best explained as a survival from earlier times when peasants
enjoyed a greater measure of independence than was normally
permitted to them in the feudal age. But villeins of this sort
dispersed over the country in substantial numbers confound
any clear-cut classification of the peasantry.

The freeman who holds in villeinage presents another ano-
maly. A man claims that he is a villein of his own accord
(*villanus sponte sua*);[3] two brothers say in court that they put
themselves into villeinage gratuitously (*gratis suis*).[4] Here the
question of tenure and status becomes very tangled. Cases arise
which show how very slender was the line which divides free
from servile tenure. Often enough it was not clear even to the
men of the time and to the men on the spot whether the tenure
was free or villein. One Simon of Sutton had disseized Cedric
'with the beard' and Saiva his wife of land in Clapham. The
jurors are certain that Cedric is not a villein, but whether he
holds freely or no, they are unable to say; however, they gave
him the benefit of the doubt and awarded him 2s. damages.[5]
Or again, a Suffolk peasant, Peter son of Ailwin, in 1203 complained that he had unjustly and without judgement been disseized of his free tenement in Thorney. The jurors admit that
he has been disseized of his tenement, but whether that tenement was free or not they do not know. Merchet had certainly
been paid when three of his sisters had been married; and
merchet in later days was supposed to be a decisive test of

---

[1] *C.R.R.* ii. 111.

[2] *Rolls of the Justices in Eyre for Lincolnshire and Worcestershire* (Selden Society,
vol. liii), no. 971. Cf. also nos. 929 and 1061, and Introd., p. xxx.

[3] *C.R.R.* iv. 234.   [4] Ibid. i. 45.   [5] Ibid. ii. 122.

villeinage. They then proceed to recite other services of a villein character which Peter had performed. Yet the matter remained in doubt; the case was adjourned in order that the highest authority in the kingdom, Geoffrey Fitz Peter, the chief justiciar himself, might be consulted on so complicated an issue.[1] One thing seems unquestionable: the free status of a man who held in villeinage did not give him the right of recovering his land by the possessory assizes. He could not claim that his land was 'a free tenement' or that he held it 'in fee'. So when in 1219 Warin son of Guy le Buis brought an assize mort d'ancestor against Nicholas le Buis concerning a bovate of land at Gunness in Lincolnshire he lost his case, for the jurors declared that, though 'Guy was a free man as regards his body and had died seized of that land, yet he held it in villeinage as did the other villeins in the same village, doing thence all villein customs'.[2] Similarly, the verdict in an assize mort d'ancestor brought by one John, son of Philip of Beaumont, was that Philip 'was not seized of that land as of fee on the day on which he died nor at any time in his life, because he held that land in villeinage as other villeins of the same village'.[3]

The free man holding in villeinage was, however, tolerably secure during his life. He could not be evicted so long as he performed his services and he could leave the holding when he pleased. These are the conclusions drawn by Bracton from a complex series of suits which began in the first year of the reign of King John and ended with a final concord in 1220.[4] In all other respects he was like other villeins. He did servile work; he might be liable to pay the degrading merchet. It was no doubt the younger sons of the free peasantry who, without other means of acquiring land, thus voluntarily entered into villeinage.[5]

[1] *The Earliest Northamptonshire Assize Rolls*, ed. D. M. Stenton (Northamptonshire Record Society, 1930), nos. 789, 793; *C.R.R.* iii. 16.

[2] *Rolls of the Justices in Eyre for Lincolnshire and Worcestershire*, no. 228.

[3] *Rolls of the Justices in Eyre for Gloucestershire, &c.* (Selden Society, vol. lix), no. 217.

[4] Bracton, ff. 199, 200; *Note Book*, pleas 70, 88. Vinogradoff discusses this litigation in *Villeinage*, pp. 78 ff. It can now be examined in the cases themselves, see *C.R.R.* i. 120, 192, 216; viii. 15, 88, 114, 216, 384; cf. also *Rot. Cur. Reg.* ii. 192. For the final concord of 1220 see *Sussex Fines* (Sussex Record Society, ii), no. 170.

[5] The emphasis sometimes placed on the position in the family (elder or younger son) suggests that the younger sons of freemen might sink into villeinage. See Selden Society, vol. liii, no. 137 and note.

But by doing so they were taking a grave risk of losing their freedom altogether. At first their free condition might be respected. But as time went on, the very obligations which they had undertaken spontaneously might be, indeed were, effectively used against them as proof of their servility. Land-owners and lawyers alike found these distinctions between different kinds of villeins a nuisance, and they worked to destroy them. In the course of the thirteenth century they tend to disappear.[1]

Nothing illustrates more forcibly the difficulties involved in determining villein status than the numerous pleas of villeinage, *placita de villeinagio* or *placita nativitatis*, as they are sometimes called. A lord *X* claims as his villein *Y*, who asserts his freedom; *A* seeks possession of his land by one of Henry II's assizes—novel disseisin or mort d'ancestor, and *B* maintains that he cannot bring such an action because he is a villein. In such cases the attitude of the courts was on the whole favourable to the weaker party; the onus of proof, it would seem, rested with the lord.[2] A jury is called upon to answer before the justices, is he a villein or not? *si sit liber homo necne?* or *utrum sit villanus?* The local jurors find the question hard to answer. The matter of status, as we have seen, first springs into importance with Henry II's legal reforms; and in the first fifty years or so there are few rules to guide the judges and the jury. There was much room for individual opinion. The common law, it has been said, was being evolved out of the common sense of the king's justices,[3] many of whom were great magnates or great churchmen, with no legal training and little experience. There was no text-book or case-book to which they might refer. It was to supply this need, to provide a manual for the guidance of the unprofessional judges, that Bracton in the middle of the thirteenth century compiled his great treatise on the laws and customs of England.[4] In this matter of villein status, however, Bracton's book is perhaps not so helpful as we would wish. He was too much influenced by the Roman law of slavery; and the serfdom of the Middle Ages is not slavery in the Roman

[1] Vinogradoff, *Growth of the Manor*, p. 345; *Villeinage*, p. 143.
[2] This is clearly brought out in a Lincoln case of 1218. Selden Society, vol. liii, no. 137 and note.
[3] *C.R.R.* i, p. v.
[4] Cf. Selden Society, vol. liii, Introd., pp. xx ff.

sense. You cannot postulate, as Bracton did, that all men are either free or serf.[1]

The chief trouble comes from mixed marriages. Freemen married bondwomen and free women married villeins. Such matrimonial intercourse shows that there were no insuperable social barriers dividing the free from the unfree. But what is the status of the progeny of such alliances? Bracton has a very elaborate set of rules governing this point; he makes freedom and serfdom depend on whether the child was the result of intercourse in the villein tenement or outside it, in a free bed (*in libero toro*).[2] But it is doubtful whether such a rule could be strictly applied. The nearest approach to a practical rule to guide the jurors in the twelfth and early thirteenth centuries was that a child followed the condition of the father. This is suggested in the compilation written in the first quarter of the twelfth century known as the *Leges Henrici Primi*. The subject is introduced in a statement of payments for murder:

'If anyone born of a servile father and a free mother is slain, he shall be paid for as a serf, because the order of birth is always derived from the father and not from the mother. If the father be free and the mother a bondwoman, the murdered offspring shall be paid for as a freeman.'[3]

So in 1206 a Norfolk man, when claimed as a villein, won his case, because the claimant could produce no kin on the father's side in evidence of servile origin, though he was able to produce two first cousins and an uncle on the mother's side who admitted that they were villeins.[4] The rule, however, that a child follows the condition of his father was not absolute. A certain William son of Simon of Stanford brought an assize of novel disseisin against William of St. Faith's, precentor of Wells, who in his turn submitted that William son of Simon could not bring the assize because he was a villein and custumer (*consuetudinarius*). The precentor was probably an unpopular man, for the clerk of the court deliberately writes his name not *Willelmus de Sancta Fide* but *Willelmus sine Fide* (Faithless). His evil reputation may have biased the court against him. For, though he was able to produce three of his opponent's kinsmen

---

[1] See *Bracton and Azo*, Selden Society, vol. viii, pp. 43 f., 49 f.   [2] f. 5.
[3] 77. 1, 2. Liebermann, *Die Gesetze der Angel-Sachsen*, i. 594.
[4] *C.R.R.* iv. 128.

(an uncle and two cousins) on the father's side who admitted that they were villeins, while William son of Simon could only produce a cousin in the third degree and an uncle on the mother's side who were alleged to be freemen, yet he, William Faithless, lost his case, and, as we find on the Pipe Roll of the same year, he had to pay a fine of three marks for making a false claim.[1]

The complexity of some of these pleas of villeinage and the thoroughness with which the justices investigated them are well illustrated by a case brought by one Aubrey of Fulburn in Cambridgeshire against his lord, Gilbert of Tany, in 1206. It is not without significance that a man in small circumstances, probably a villein, presumes to embark on a lawsuit with a feudal magnate, the owner of an Essex barony of $7\frac{1}{2}$ fees, who was shortly to be sheriff of that county, and who held besides a considerable property in Cambridgeshire. Aubrey complains that Gilbert imprisoned him when he carried the king's writ into his court, he being a freeman and holding freely. Gilbert defends the imprisonment on the ground that Aubrey is his villein. The elaborate family tree of Aubrey is remarkable, and would be the envy of the modern pedigree hunter.[2] No less than thirty of his kinsmen are named. Gilbert of Tany actually produced in court nine of Aubrey's relatives who all declared that they were villeins, six on the father's and three on the mother's side. Aubrey produced four who were admitted to be freemen. Here three mixed marriages are in question to complicate the issue. Aubrey proffers half a mark for an inquest and a jury of the neighbourhood who should decide *utrum liber sit necne*. For one reason or another the case was continually adjourned; it is nine times before the court; and when in 1211, six years after it was opened, we lose sight of it, it was still undecided whether Aubrey was a freeman or a villein.[3]

I will cite one more case which is of interest from the point of view of the defence put up by the peasant. 'Werric de Marines seeks Ralph son of Segar as his villein, as him who was the son of Segar who himself died as his villein', and he pro-

---

[1] *C.R.R.* iv. 195; *P.R. 8 Jo.*, p. 34.

[2] This and a number of other pedigrees of villeins have been conveniently set out by Miss Cam, *Liberties and Communities in Medieval England*, ch. viii.

[3] *C.R.R.* iv. 305.

duces five near relatives who declare in court that they are villeins. Ralph does not deny that these men are his kinsmen, but he takes exception to them as evidence of his condition. Thus the witness of one of them, Seman, a great-uncle, ought not to be accepted because he had 'intruded himself into villeinage'; he was, that is to say, a freeman who was holding in villeinage, and not a villein by birth. He objects to another, Ralph son of Lece, on the ground that Lece's parents were not properly married, and therefore Lece herself was illegitimate. Two others, he says, should not be produced in evidence against him because they are kin on his mother's side. He then brought out his trump card, a charter of Hugh de Marines, brother of Werric, which testified that this Hugh had granted him one virgate of land in Westmill to be held of himself and his heirs by the service of half a mark annually in lieu of all services. Both parties proffer money for an inquest.[1] Werric's proffer of 1 mark is entered on the Fine Roll and accounted for on the Pipe Roll.[2] A jury of knights and other law-worthy men is summoned to appear to examine the truth of the matter. Ralph's case apparently was not so good as he had at first supposed. At all events, on the day fixed for the hearing he neither appeared nor excused himself, and the case went against him by default. However, to avoid any risk of the reopening of the question in the future the parties deemed it advisable to have the decision of the court put on record on the rolls of the Exchequer.[3]

The case of Ralph, son of Segar, raises the problem of commutation of services for a money rent. He produced in court a charter (perhaps a forged charter, for, as we have seen, he lost his case) which purported to grant a virgate of land for the render of half a mark a year in lieu of all services. We are apt to regard this question of commutation as one of significance only in the fourteenth century. It was in fact a prominent one in the twelfth; and it is very relevant to the question of the status of peasants. If the peculiar mark of villeinage is the performance of labour services, and particularly uncertain labour services, are we justified in regarding a man as a villein

[1] Ibid. 22, 54.
[2] *Rot. de Ob. et Fin.* 247, 277; *P.R.* 7 *Jo.*, p. 195.
[3] *P.R. 8 Jo.*, p. 235.

who, instead of these indefinite services, pays a definite sum of money by way of rent?

In an important paper recently contributed to the Royal Historical Society,[1] Professor Postan drew attention to the fact that in nearly all the surveys which belong to the twelfth century, or which reflect twelfth-century conditions, there are clear indications of a widespread movement in the direction of commutation of services for rents. There is a large and growing class of tenants who render no services or only insignificant services, but instead pay a money rent. The evidence for this is not confined to one particular locality but extends to estates distributed over every part of England. We find it on the estates of the bishop of Durham in the north, and on those of the abbot of Glastonbury or the nuns of Shaftesbury in the south-west; we find it on the manors of the abbey of Ramsey in the east and in the villages of Staffordshire and the adjacent counties belonging to the abbey of Burton; it is evident again on the estates in the Home Counties belonging to the dean and chapter of St. Paul's and on those belonging to the bishopric of Worcester in the south midlands; the same is true of the scattered manors owned by the Knights Templars. Perhaps the process of development can be seen most clearly on the English estates of the abbey of the Holy Trinity at Caen. These were surveyed at three periods in the course of the twelfth century: in the early years of the reign of Henry I, in the beginning and again towards the end of the reign of Henry II. Thus on the manor of Minchinhampton in Gloucestershire in the first survey 26 virgates were held by service and 9 by the payment of rent; in the second, 9 were held by services, 11 were wholly commuted, and 11 partially commuted (that is to say, with the option of performing services or paying rents); in the last survey all the virgates were commuted.[2]

We are not here concerned with the economic causes which produced this change. In the case of the nuns of Holy Trinity at Caen it was an obvious convenience to manage their distant overseas estates on a monetary basis. In other cases a temporary agricultural depression may have induced these ecclesiastical

[1] *Transactions*, Fourth Series, xx (1937), 169.
[2] Ibid., p. 183.

landlords to lease their demesne lands and take rents from the villeins. This was probably so on the manors belonging to Burton abbey. Two surveys were made of these manors, both in the reign of Henry I, and perhaps less than a dozen years separate them. Yet in this short space of time great changes have taken place. There is some fluctuation between work and rent; so, for instance, at Mickleover in Derbyshire 'Edwin, who was the man of Orderic, has two bovates by work (*ad opus*) which previously Lewin had by rent (*ad malam*)'; while 'William son of Ernold has two bovates for two shillings which previously Osmer held for work'.[1] But the general trend is clearly in the direction of commutation of work for rent. At Abbots Bromley in Staffordshire all the villeins, who had been rendering two days' week-work besides other customary services, have in the interval changed over to rents. Indeed, four of the peasants on this manor have together with the priest taken a lease of the whole demesne for twenty years at a rent of 100*s*. The abbot has relinquished all his farming activities at Abbots Bromley.[2]

These rent-paying peasants are termed *censarii*, or sometimes *molmen* or *firmarii*. On one of the Burton manors, that of Appleby on the borders of Derbyshire and Leicestershire, they are sharply distinguished. There are *puri villani* and *puri censarii*.[3] Now if, as Bracton maintains, the mark of a villein is the uncertainty of his services, what is the position of the peasant who has commuted these so-called uncertain services for a fixed rent? The rent might be raised or lowered from time to time,[4] but its payment cannot be reckoned an uncertain service in the sense in which Bracton uses the term, that he does not know in the evening what he will have to do on the morrow. Yet in fact these rent-paying peasants were still at the lord's mercy. Even in these early surveys there are a few instances, as we have seen, of a reversion from rent to labour. The tenants

---

[1] The *Burton Cartulary*, ed. Bridgeman, in the Transactions of the William Salt Arch. Society, 1916, p. 230.

[2] Ibid., pp. 222–3. The rent of the demesne was subsequently raised by 20*s*., when the abbot gave up the wood which he had at first reserved from the lease.

[3] Ibid., pp. 244–5.

[4] There is evidence of a general rise in rents in the interval between the two surveys.

at Abbots Bromley, however, were left undisturbed on their rented holdings for over a hundred years.[1] Then, in 1236, the abbot, impressed perhaps by the agricultural prosperity which then prevailed, tried to reclaim the labour services of his tenants. Bromley was ancient demesne of the Crown,[2] and the peasants sought and obtained a writ *monstraverunt*:

'Henry by the grace of God &c. to the abbot of Burton greeting. Your men of Bromley have shown [*monstraverunt*] us that you exact from them other customs and other services than they ought to do and were accustomed to do in the time when the manor of Bromley was in the hands of our predecessors, kings of England. And therefore we command you that you do not exact other customs or other services than they ought to do and were accustomed to do in the said time. And unless you do this at our command, we shall order it to be done by our sheriff of Staffordshire. Witness myself at Kempton on the 20th day of October in the twentieth year of our reign [1236].'[3]

If the documents relating to this litigation were correctly copied into the cartulary of the abbey (and there is no apparent reason to doubt it), this is an early, if not the earliest, known example of the precise phrasing of this writ, which was designed to protect the privileged position of villeins on ancient demesne.[4] In this case, however, their special remedy availed them nothing. The abbot Laurence refused to comply with the king's demand; he busied himself actively in the defence of what he maintained were the rights of the abbey 'not without great labour and

---

[1] Abbots Bromley was made a borough in 1222. But the terms of the charter (*The Burton Cartulary*, ed. Wrottesley for the William Salt Arch. Society, v, pt. 1, 1884, p. 73) show that it retained its essentially rural character, and the *censarii* cannot have been full recipients of the privileges. Cf. Miss Deanesley's preface to the *Catalogue of Charters and Muniments belonging to the Marquis of Anglesey* (Staffordshire Record Society, 1937), p. xxxviii: 'The burgesses had lately been the abbot's censarii, and though their status was raised by the grant, traces of the original tenure remained.'

[2] *D.B.* i. 246 *b*.

[3] The *Burton Cartulary*, ed. Wrottesley, p. 65. Henry III was at Kempton on 21 October 1236: *Close Rolls 1234–1237*, p. 381.

[4] It is earlier than any case cited by Pollock and Maitland, *Hist. of Eng. Law*, i. 389 n. H. G. Richardson and G. O. Sayles in *Select Cases of Procedure without Writ* (Selden Society, vol. lx, p. xcv, notes 1, 5) in fact doubt Maitland's view that 'such writs were in use early in Henry III's reign'. Commenting on a case relating to the disabilities of tenants in villeinage which was heard in 1258, they say: 'At a later date they might have obtained a *monstraverunt*.'

expense'. He sought out the king at Marlborough and again at Kempton,[1] and succeeded by means of a writ *pone* in getting the suit transferred to Westminster. But it dragged on for seventeen years before a verdict was given against the peasants at Nottingham in 1253. They were adjudged still to be villeins on the ground that they paid merchet for marrying their daughters, they paid tallage (or *stud*, as it is called in the record) at the will of the abbot, and owed villein services.[2]

A case of a similar kind (but without the additional complication of ancient demesne) came before the king's court in 1214. A Northamptonshire peasant brought an assize of novel disseisin; he claimed that certain persons had unjustly and without the judgement of a court deprived him of his free tenement. The jurors found that he and his father before him had held that land by customary services, by fork and flail, and that they could not give a daughter in marriage without payment, but the father had arranged that the works and customary services should be commuted for money (*ponebantur ad denarios*) as long as he pleased; and in this way he had held the land for twenty years. But the son had no claim on the protection of the court; he was still a villein.[3]

The commutation of services for a money rent did not then alter a man's status. He remained a villein, and his services could be reclaimed. Nevertheless, there are some indications that suggest that he was regarded as freer than he had been by reason of his commutation. Henry of Blois, bishop of Winchester and brother of King Stephen, was also abbot of Glastonbury from 1126 till his death in 1171. During his abbacy he made many changes in the administration of the Glastonbury manors. These changes are reflected in an inquest of these estates made by one of his successors, Abbot Henry de Sully, in 1189. One of the questions put to the jurors was 'if any land, which ought to render work, had been made free in the time of Bishop Henry or since, and by what authority this was done and to what extent is it free'. Of a certain Anderd Budde it is said that he 'holds more freely than his predecessors

---

[1] The king was at Marlborough 7–15 December 1236, and at Kempton 2–10 February 1237.

[2] *The Burton Cartulary*, ed. Wrottesley, p. 66.

[3] *C.R.R.* vii. 60–1.

were accustomed to hold'.[1] Evidently commutation affected in some way the condition of the tenant. There were degrees of freedom. The Glastonbury estates were again surveyed twice in the thirteenth century; and these reveal, as on the Burton manors, a full return to the system of labour services.

Commutation is not an equivalent to enfranchisement. It required more than a mere change in obligations to be rid of serfdom. Many a villein no doubt gained his liberty by the well-known method of escaping to a town where, if he remained unclaimed for a year and a day, he became free. The German proverb *Stadtluft macht frei* (Town air enfranchises) applies to England no less than to the Continent. But there was the risk of recapture and punishment. When we read on the Pipe Roll of 1182 that Richard son of Waldeve 'renders account for £5 for having right over his men who made themselves free when they are not',[2] we probably have to do with recaptured villeins who sought their liberty in this way.

It is generally assumed that a villein obtained his freedom by entering the orders of the Church. He could only do this, at least after 1164, with his lord's consent, for the latter was losing thereby a piece of his property. It would seem, however, that this restriction was not entirely confined to peasants. In 1196 a certain Otto of Scarborough, who certainly was not a villein, became a monk at the priory of Llantony near Gloucester. On the Pipe Roll of that year[3] there is an entry that Otto owes 20 marks—a big sum—for licence to enter religion. He may already have made over all his worldly goods to the priory, for four years later demand for the payment of Otto's debt of 20 marks is made of the prior himself.[4] The requirement of the consent of the lord, therefore, proves little. The question is, did admission to the Church, like admission to the town, necessarily carry with it enfranchisement? Bracton implies at least conditional enfranchisement when he says that a serf, who has become free by entering the Church, becomes a serf again if he returns to secular life.[5]

The condition of the lower clergy at this time is far from

[1] *Liber Henrici de Soliaco, Abbatis Glaston., 1189*, ed. J. E. Jackson (Roxburghe Club), 1882, pp. 21, 121; Vinogradoff, *Villeinage*, pp. 168-9.
[2] *P.R. 28 Hen. II*, p. 63.   [3] *8 Ric. I*, p. 187.
[4] *Memoranda Roll 1 Jo.*, p. 32.
[5] f. 5. Cf. *Bracton and Azo* (Selden Society, vol. viii), p. 51.

clear. The parish priest would certainly not rank among the
aristocracy of the medieval village. He was often of humble
birth, poorly educated, and seldom rich. He had his solemn
duties to perform, which no doubt enhanced his position; he
baptized, married, and buried the peasants; sometimes he
preached to them. But on week-days he was working, like any
other peasant, in the fields. He associated with villeins, and
often seems to be a villein himself. In many Domesday entries
the priest is joined with the villeins; there are so many villeins
and a priest.[1] At Barksdon Green in Hertfordshire two villeins
with the priest and five bordars have two ploughs among them.[2]
In the Cambridgeshire Inquest, which is believed to be a copy
of the original returns, it is stated that at Kennett there was
a priest and six villeins; in the parallel passage in the Exchequer
Domesday there is mention of seven villeins but no priest.[3]
These entries become more striking if they are contrasted with
the precise description of Aluric, who before the Conquest held
land at Horndon-on-the-Hill (Essex) as *presbyter liber homo*,[4] or
of Edwin, who in the time of King Edward held three virgates
at East Ham, as *liber presbyter*.[5] In a twelfth-century list of the
villagers of Wetmoor in Staffordshire Ailwin the priest appears
after five villeins and before the cowman; and precisely like the
cowman, he has his house and a croft, and does one day's work
a week for his lord.[6] If they were not villeins, the parish priests
often held in villeinage. In the *Red Book* of Worcester, a survey
of the estates belonging to Worcester priory in 1182, there is
a list of the tenants of the manor of Northwick, the present
village of Claines. Against the name of each tenant there is
a note stating whether he holds freely or in villeinage. There
are two priests; one is said to hold freely, the other in villeinage.[7]
The well-known story of Simon of Elmham, prior of Norwich,
illustrates the disadvantageous position of the priest of servile
origin. It might stand in the way of his promotion. In 1236

---

[1] Cf. the passages collected by J. H. Round in *V.C.H. Essex*, i. 384.
[2] *D.B.* i. 137.
[3] Ed. N. E. S. A. Hamilton for the Royal Society of Literature, p. 1. *D.B.* i.
196 b. See A. Ballard, *The Domesday Inquest*, p. 189.
[4] *D.B.* ii. 42.   [5] Ibid. 64 *b*.
[6] *Burton Cartulary*, ed. Bridgeman, p. 220. Ailwin also rented two bovates for
3*s*. (ibid.).
[7] Ed. M. Hollings, *Worc. Hist. Soc.* (1934), p. 31 f.

he was elected bishop of Norwich. His election was subsequently quashed by the pope and the king on what Matthew
Paris calls 'some ridiculous grounds'.[1] What these grounds were
we know from an entry on the Liberate Roll of 1237, which
contains an instruction to the custodian of the bishopric to
cause five witnesses to come before the legate, men 'who best
know and will prove the prior of Norwich to be of servile
condition'.[2]

The town and the Church were, so to speak, bolt-holes
through which villeins might or might not succeed in escaping
and ultimately gaining their freedom. Manumission was a
solemn and definite act.

The only sure methods of obtaining enfranchisement were
by charter or by purchase. In the latter case the purchase
money must be provided by a third party, as the villein's property was in theory his lord's. Thus a Lincolnshire woman,
when captured and imprisoned as an escaped villein, claimed
that she had been redeemed by another and by the money of
another (*redempta per alium et per alterius pecuniam*).[3] But there
would not have been any permanent record of such a transaction. A charter was safer and more satisfactory.

At Allerthorpe, an estate of the Knights Templars in the East
Riding of Yorkshire, a tenant named Arngar held two bovates
for 8s. in lieu of all service, and, it is added in the inquest taken
in 1185, 'he was accustomed to do service, but by the charter
of the Master of the Temple he was made free'.[4] Ralph the
Priest gave his villein tenement of 20 acres to the Augustinian
priory of Coxford, and, when his lord William of Pinkney complained, Ralph produced charters to prove that he held freely;
and as William could not deny the charters, Ralph won his
case.[5] Few of these early charters of enfranchisement have
survived. One, which evidently belongs to the later years
of the twelfth century or the early years of the thirteenth, is
worth quoting because it illustrates not only the form in which
such charters were drawn up, but also that the method of

[1] *Chron. Maj.* iii. 389.
[2] *Cal. of Liberate Rolls, 1226–40*, p. 299.   [3] *C.R.R.* v. 77.
[4] *Records of the Templars in England in the Twelfth Century*, ed. B. A. Lees, p. 124.
[5] *C.R.R.* v. 94. That William of Pinkney is claiming that Ralph was his villein
and not merely holding his land in villeinage is clear from the fact that he brings
into his evidence that he had sold Ralph's sister for 4s.

granting emancipation by charter was sometimes, perhaps usually, combined with the other method, that of purchase:

'Know all both present and future that I, Emma of Dummart, with the consent of my heirs have liberated William my native, the son of Baldwin; and I have granted to him and by this my charter have confirmed that he be free of all servitude to come and to go freely and quit, as a freeman, wherever he wishes. And for this liberty and confirmation Richard son of Hugh has given to me for him fifteen shillings of silver. And if anyone shall challenge him, I and my heirs will warrant him against all men. And I Engelram of Dummart at the request of the friends of the said William have confirmed to him this enfranchisement and ratified it by the impress of my seal.'[1]

It will be seen that it is not William the villein but another man, Richard son of Hugh, who pays the money. The rule, therefore, that a man cannot purchase his freedom with his own money is complied with.

If a liberated serf could not find the capital sum necessary to purchase his freedom he might arrange to pay an annual quit-rent. In the early years of the thirteenth century one Alan son of Morice of Wlstorp was 'redeemed from the yoke of servitude' by the hand of the custodian of the house of St. Barbara of Colsterworth (Lincolnshire). The terms of his enfranchisement were as follows:

'I, the said Alan, as long as I live will pay each year to the said custodian of Colsterworth or his successors six pence at two terms of the year, that is at Christmas threepence and at the feast of St. Michael threepence, and I will be obedient and tractable.'[2]

Sometimes freedom was granted not in return for cash but for services rendered, for example going on crusade in place of

---

[1] Madox, *Formulare*, p. 417. Emma of Dummart and her sister Aliz succeeded their brother Engelram at Oxhill, Warwickshire, as joint heiresses in 1186 (*P.R. 32 Hen. II*, p. 133. Cf. *Rot. de Dominabus*, Pipe Roll Soc. xxxv, p. 23). They are entered on the Pipe Rolls as owing scutage in the first four years of the reign of King John (*P.R. 1 Jo.*, p. 254; *4 Jo.*, pp. 37, 40). Oxhill was in the hands of the crown in 1216, when the king gave it to Theodoric of Whichford (*Rot. Lit. Claus.* i. 289). Cf. Dugdale, *Antiquities of Warwickshire*, pp. 429–30.

[2] The charter is preserved among the MSS. of Eton College: *Eton College Records, vol. 8, Documents relating to the Alien Priory of Beckford, co. Glouc., and Colsterworth, co. Lincs.* no. 1. My thanks are due to the Provost and Fellows of Eton for permission to print it as an appendix to the present chapter (p. 34). For the transcript I am indebted to Mr. Noel Blakiston of the Public Record Office.

his lord. Professor Stenton has printed a series of charters by which William of Staunton, Nottinghamshire, so enfranchised his serf *qui pro me ibit in sanctam terram*. This was on the third crusade, and he endowed him with two bovates of land in return for a yearly render of a pound of incense and a pound of cummin to the rector of the church.[1] The crusade, involving the absence of many lords from their lands, doubtless worked a social upheaval. When the great Lincolnshire baron Guy of Craon left for the East in the army of Richard I, he converted a servile into a free tenure for a period of four years, with the further stipulation that, if he did not return within the four years, the land should remain free and quit to the tenant and his heirs. The arrangement came to nothing, for Guy returned safely.[2] But it illustrates how the prospects of peasants might be improved by the departure of their lords on the great adventure.

The enfranchisement of a serf was not only a definite act; it was also a solemn and a public act. It must be made known to all that the villein has been released from his bondage. According to the author of the *Leges Henrici Primi*, he who liberates his serf shall do so in church or market, in the county or hundred court, openly in the presence of witnesses, and he shall show him the ways and open gates, and shall place in his hands the lance and sword or whatever are the arms of freemen.[3] The *Leges* are a strange jumble of archaic custom and contemporary law; they are taken from Anglo-Saxon dooms, from continental sources, and from Anglo-Norman custom. The part of this text about the free ways and open gates is borrowed almost verbally from the *Lex Ribuaria*, the law of the Ribuarian Franks, which dates from the second half of the sixth century, and has to do with the enfranchisement of the slave in the Roman sense.[4] But the rest of the passage seems to relate to Anglo-Norman custom, and is indeed repeated in a compilation made at London in the time of King John.[5] In this later text there is a significant addition. The lord shall present the

---

[1] *Eng. Hist. Rev.* xxvi (1911), 93 ff.

[2] *C.R.R.* ii. 13 f. The tenant in 1201 attempted to benefit under this agreement, but William of Longchamp, Guy's son-in-law and heir, had no difficulty in proving his right on a plea of villeinage.

[3] 78. 1; Liebermann, op. cit., i. 594.

[4] c. 61. 1. *Mon. Germ. Hist. Leges*, v. 252.

[5] *Willelmi Articuli Londoniis Retractati*, 15; Liebermann, op. cit. i. 491.

liberated serf to the sheriff in full county court, taking him by
the right hand and there claim him quit from the yoke of
servitude by manumission.

The elaborate ceremony described in these early law-books
is not recited in any of the legal records of this time. The quit-
claim by the lord is, however, stated by Glanvill as the first of
the several ways by which a serf could be brought to freedom.[1]
In this way two Cambridgeshire peasants obtained their liberty
in 1203. The entry on the Plea Roll notes that these men who
had been accused of being villeins by the steward of William
of Colevill remained free of their bondage by the acknowledgement (*per recognitionem*) of William of Colevill in the presence
of the justices of the bench.[2] Declaration in the county court
was the method by which solemn acts affecting a man's condition were given publicity. A sentence of outlawry, for example,
must be pronounced in the county court.

The investiture of the liberated serf with the arms of a freeman, the spear and lance, is also in keeping with what we know
of Anglo-Norman law and custom. The villein of the twelfth
century was not trusted to carry weapons. In the Assize of
Arms of 1181 he is expressly excluded. 'No one', it declares,
'shall be received to the oath of arms except a freeman.'[3] There
is no hint that he was capable of bearing arms until 1225, when
the villeins' arms are excepted from assessment of a tax on personal property.[4] In 1198 a man, claimed as a villein by the
abbot of Evesham, protested his freedom on the ground that
on the assize of the lord king he was sworn as a free man to
have arms.[5] The right and privilege of carrying weapons was
therefore one of the distinctive marks of free status, and originally the bestowal of these arms was part of the formal ceremony
of manumission.

The bestowal of arms on the liberated serf has its parallel
in the ceremony of knighthood or the creation of an earl, in
which the girding on of the sword was an essential feature.
When in 1200, after his divorce of Isabel, countess of Gloucester,
King John granted the earldom of Gloucester to Almaric,
formerly count of Evreux, he wrote to the justiciar, Geoffrey
Fitz Peter, informing him that 'we have girded him with the

---

[1] v. 5.   [2] *C.R.R.* iii. 39.   [3] Cap. ix (Houeden, ii. 263).
[4] *Foedera* (Record Commission), i. 177.   [5] *C.R.R.* i. 45, 67.

sword and then made him an earl' (*ipsum gladio cinximus et comitem inde fecimus*).[1] Indeed, the sword plays a prominent part in all ceremonial investitures up to the highest of all, the coronation of the king. The right to carry arms gave a dignity to the peasant which put him more on an equality with men of rank. It was perhaps the most highly prized privilege of the liberated peasant.

[1] *Liberate Roll 2 Jo.* (Pipe Roll Soc., N.S. xxi), p. 89. For the circumstances in which the count of Evreux became earl of Gloucester see *Complete Peerage*, v. 692. Cf. also King John's writ of 28 December 1201, 'Sciatis quod nos concessimus Roberto de Leveland, cui arma dedimus die Circumcisionis Domini' (*Rot. de Liberate*, ed. Duffus Hardy, p. 25), which presumably refers to the ceremony of knighthood.

APPENDIX (*see p. 31*)

'Noverint omnes hoc scriptum visuri vel audituri quod ego Alanus filius Moricii de Wlstorp redemptus a jugo servitutis per manum custodis domus Sancte Barbare de Colsteword homagium et fidelitatem eidem custodi et ejus successoribus feci et securitatem fide media et per bonos et legales homines fidejussores meos scilicet Amfridum de Ultra Aqua Hugonem de Muney fratrem meum Robertum Canterel Radulfum Coutte qui omnes fidem dederunt pro me in solidum quod ego dictus Alanus quamdiu vixero singulis annis solvam predicto custodi de Colstword vel successoribus suis vi denarios ad duos anni terminos scilicet ad Natale domini tres denarios et ad festum Sancti Michaelis tres denarios et obediens et tractabilis ero. In hujus rei testimonium sigillum meum huic scripto apposui.'

III

THE KNIGHTS

IN considering feudal society one inevitably thinks of knights and castles. As Madox, the great eighteenth-century antiquary, puts it, 'the men covered with steel domineer over burgesses and peasants; the armed over the unarmed'.[1] This is obviously true; the knights play a very active and a very prominent part in the life of the twelfth century. But the statement is liable to give a misleading impression. They were never a numerous class in this period. A man was not a knight until he was of age[2] and had been made so by a formal ceremony. When we hear of this ceremony of knighthood it is usually a great man, a king's son, an earl, or a great baron who is dubbed.[3] The knighting of a humbler person is seldom noticed. It is not even clear who were qualified for this dignity. In 1224 it was a man of full age who held an integral knight's fee or more that was required 'to take up arms and cause himself to be made a knight'.[4] It is, however, evident that some years earlier men who held less than a whole fee, if they were otherwise qualified, were expected to take up the status. Thus a return of an inquest concerning the military and free tenants of the see of Durham, when the bishopric was vacant and in the hands of the Crown between 1208 and 1210, records that 'Jordan Ridel holds Tilmouth (Northumberland) and he does the service of half a knight'. Against this entry the clerk has noted 'Let him come and be knighted' (veniat et fiat miles).[5] There was already a tendency, which becomes very marked a few years later,[6] among men duly qualified for the position, to seek to escape from the responsibilities which knighthood

---

[1] *Baronia Anglica*, p. 19.

[2] Walter de Haule was taken into custody because he gave his sister to William de Bodiham and made him a knight *dum fuit infra etatem* (*C.R.R.* vi. 54). In 1220 Henry of Tracy claimed the manor of Barnstaple which, he said, 'King John rendered to him as his inheritance when he made him a knight' (ibid. viii. 365).

[3] The most elaborate account of the ceremony of knighthood is perhaps that of the knighting of Geoffrey of Anjou by Henry I at Rouen in 1129 (*Chroniques des Comtes d'Anjou*, ed. Halphen and Poupardin (1913), pp. 178 ff.).

[4] *Rot. Lit. Claus.* ii. 69 b.   [5] *Book of Fees*, p. 27.

[6] See *P.R. 14 Hen. III*, Introd., p. xxii, where instances of the fine *ne fiat miles* are collected.

demanded. Men would often prefer to pay a fine and remain
mere country gentlemen. On the other hand, there were some
thousands owing knight service, or more particularly fractions
of knight service, who were not knights or capable of becoming
knights. They were owners of small properties whose sole
obligation in respect of knight service was a contribution of
a few shillings of scutage when scutage was due. Walter de
Becco, for example, was charged (and paid) 20s. for the third
scutage of Richard I, assessed in 1196 at 20s. on the fee.[1]
Walter, therefore, owed the service of one knight. But it is
recorded on the Memoranda Roll of 1199 that 'the sheriff says
that he holds in free socage' (*tenet de libero socagio*).[2] There were
many men, like Walter de Becco, holding by knight service
and paying scutage, who in fact belonged to that large and
ill-defined class which came to be generally known as tenants
in free socage.

In these circumstances it is impossible to arrive at anything
approaching a precise estimate of the actual number of knights.
In Round's view it 'can scarcely have exceeded, if indeed it
reached, 5000 knights'.[3] This is almost certainly an under-
estimate. But it would be safe to put the figure at between
6,000 and 7,000, which in a population of something under
three million is not a large proportion. They form, therefore,
a relatively small group in the structure of society. Nevertheless,
their importance, their prominence, their power to dominate
was at the height of the Middle Ages quite incommensurate
with their numbers. They are the principal element in the
feudal conception of society.

The essential features of knight service have long been established. The groundwork of Round and Maitland in the last
generation has been further developed and illustrated by
scholars of our own day. There is, however, still room for
discussion on points of detail. We have always to bear in mind
that it is a rapidly changing age, and what is true of the beginning of the twelfth century is not necessarily true of the end of
it. The social outlook has changed; the art of war has changed.
The army with which William the Conqueror defeated Harold

---

[1] *Chancellor's Roll, 8 Ric. I*, p. 122; *P.R. 3 Jo.*, p. 132.
[2] Pipe Roll Society, N.S., vol. xxi, p. 78. This is a very early instance of the
use of the phrase *liberum socagium*.   [3] *Feudal England*, p. 292.

at Hastings would be almost as unserviceable for the wars waged by Richard I against Philip Augustus as the armies of Wellington would be for fighting the Germans to-day. When William I parcelled out the land of England among his companions in return for the service of knights, he expected to be provided with a small force of mounted warriors specially trained to the profession of arms, always ready to turn out for a short period of service when needed. The requirements of Richard I were very different. He had to prepare for campaigns which might last for a long time; in some years war was almost continuous, and very much larger armies were necessary. To meet these new developments the feudal levy had to be reconstituted.

In the early days of knight service the knight has two military functions. He must fight in the army when there is fighting to be done; and whether there be war or peace, he may be required to garrison royal castles. These two duties, though overlapping, are distinct. Let us consider first his place in the army. The knights formed the core, the rank and file, of the feudal army. 'Knighthood in the eleventh and early twelfth centuries', writes a leading authority on the subject,[1] 'had denoted nothing beyond proficiency in the art of fighting on horseback.' This emphasis on cavalry fighting may perhaps be exaggerated. In all accounts of medieval warfare we hear a good deal about knights charging about on their horses, often rather ineffectively. But a castle, the mainstay of medieval defence, could not easily be captured by a cavalry charge; and even in this early period the importance of infantry as giving stability to the line of battle seems to have been appreciated. It was not so much his horse as his professional training and his better arms and better equipment which gave the knight his superiority in fighting. In the battle of Tinchebrai, which won Normandy for England in 1106, and again at Bremûle in 1119, King Henry I himself fought on foot among a number of dismounted knights. At the battle of the Standard in 1138, where the army of King David of Scotland was destroyed, the Anglo-Norman knights fought on foot; and again the final phase of the battle of Lincoln in 1141 was a furious attack on the reserve of dismounted knights commanded by King Stephen, himself

---

[1] F. M. Stenton, in *History*, xix (1935), 298.

on foot. Nevertheless, these striking facts should not obscure the prominent part played in early medieval warfare by companies of mounted knights under the command of the king himself or of one of his greater barons. It was the characteristic feature.

We may suppose that in the early days, when knight service was first instituted, the normal campaign was expected to last for a few weeks in the summer months. The campaign of Tinchebrai, for instance, occupied the months of August and September. Winter campaigning was exceptional, and it was usual to make a truce in the autumn to last until the next Easter. It was a short term of service that was demanded from a knight. How short this period was is not certainly known, but it has generally been assumed to be forty days.[1] The grounds usually given for this assumption are, first, that it appears to have been the practice in the Norman duchy, and secondly, that when the service was commuted for money, the sum was assessed in terms of a knight's wages for such a period. It may be added that forty days was a recognized period of service in other conditions. It was, for example, very commonly the period for which knights were hired for garrison duties. Thus the castle of Kenilworth was garrisoned in 1193 by 5 knights and 10 mounted sergeants for forty days; that of New-castle-under-Lyme in the same year by 5 knights, 15 mounted and 30 foot sergeants for the same period.[2] Similarly, the term of service required from military sergeanties was very commonly forty days. Thus Thomas Buffin, who held a small estate at Nethercote in Oxfordshire, was required to be 'in the army of the lord king for forty days at his own cost, and if he shall be in the army for more than forty days, it shall be at the king's cost'.[3] Such a precise statement is in strong contrast with the vague expressions about knight service. In neither of the two great books kept at the Exchequer for reference about services and precedents, the *Red Book of the Exchequer* and the *Book of Fees*, is there any mention of the term of service demanded. The statement is invariably that So-and-so owes the service of one knight, of two knights, of half a knight, or whatever it may be.

[1] Maitland states this with the caution that it 'can hardly be proved for England out of any authoritative document' (Pollock and Maitland, *Hist. of Eng. Law*, i. 254).
[2] *P.R. 5 Ric. I*, pp. 57, 83.   [3] *Book of Fees*, pp. 830, 1375.

Some light is thrown on this little problem of knight service by a charter which, though printed as long ago as 1835,[1] has not been discussed in this connexion. It is a grant by John Fitz Gilbert, the father of William Marshal, who flourished between 1130 and 1164, to Hugh of Ralegh, of lands in Nettlecombe in Somerset. These lands were to be held for the service of one knight in such a way that 'if there is war, he shall find for me one knight for two months, and if there is peace for forty days for such service as knights of barons ought reasonably to perform'. This suggests that the knight was expected in time of peace to spend a period of forty days in the year in training, perhaps in the garrison of a castle, just as the territorial soldier of modern times has to put in a fortnight in camp. But in time of war a longer period of service was demanded. Two months might suffice for the military needs of the eleventh and early twelfth centuries, but not of a century later. The king had then to fall back on the expedient of calling up a quota of the military service due to him instead of the whole. But these few must be prepared to serve for the duration of the war.

The feudal levy must very soon after its institution have revealed its weak points. The greater barons with their retinues of knights, often trained and maintained in their households, could no doubt put into the field a contingent of good fighting quality. But the smaller tenants must have often been in difficulties. There was no provision for the case of knights who through age or infirmity were prevented from fighting. And the position of the ecclesiastical tenants, the bishops and abbots, who among them, according to the returns made in 1166, were responsible for more than 750 knights,[2] was far from easy. One of the minor quarrels which William Rufus had with Archbishop Anselm arose from the fact that the Canterbury contingent sent to the Welsh war in 1097 were not properly trained and of poor quality.[3] In the early days of feudalism it was common for the ecclesiastical, like the secular tenants-in-chief, to retain knights about them to carry out their military obligations. Domesday, for instance, records twenty-five houses in

---

[1] In *Collectanea Topographica et Genealogica*, ed. Madden, Bandinel, and J. G. Nichols, ii. 163 f. I owe this reference to the kindness of Professor F. M. Stenton.

[2] See the table in H. M. Chew, *Ecclesiastical Tenants-in-Chief and Knight Service*, pp. 4–5.

[3] Eadmer, *Hist. Novorum*, ed. Rule, p. 78.

the neighbourhood of Westminster which were set apart for the
abbot's knights and other men; at Ely these lived in the *aula
ecclesiae* and fed at the abbot's table.[1] But the presence of a
body of irresponsible and insubordinate men-at-arms about the
place was found to disturb the even tenor of monastic life, and
it was soon abandoned in favour of the practice of planting
them on the land. That is to say, the lord contracted with
a tenant for the performance of the military service in return
for a landed estate. But the nature of these contracts between
lord and tenant were, as we shall presently see, often elaborate
and they were certainly difficult to enforce.

A further cause of confusion arose from the practice adopted
by many tenants-in-chief of enfeoffing more knights than were
required to fulfil their military obligations. How in such cases
was the service to be apportioned? Some service other than
military service must have been assigned to the superfluous
knights, who obviously did not get their lands for nothing.
They may have assisted by contributions of money. The four
knights of the abbey of Ramsey who were in the king's service
in 1212 were to be supported by the knights of the abbey who
remained at home.[2] This was probably the general practice.
But sometimes there seems to have been an element of uncertainty. A curious case comes from Winchester. The bishop
owed the service of sixty knights; he had enfeoffed seventy.
One of these was Nigel of Broc, who was so vague as to what
was expected of him that in 1201 he put himself on the grand
assize in order that a jury might decide 'whether he owed the
bishop the service of one knight for the fee which he held of
him in Braishfield, or whether he should rise for the bishop in
the king's court and make room for him so that he can speak
with the lord king'.[3] It was this sort of confusion and the
difficulty of enforcing services that accounts for the general
popularity of the system of commutation of the service for
money.

It is needless here to enter into the question of the antiquity
of scutage. Suffice it to say that it appears almost as soon as
the introduction of knight service into England, and that a writ

[1] J. Armitage Robinson, *Gilbert Crispin*, p. 40 f.; Chew, op. cit., pp. 114 ff.
[2] *Rot. Lit. Claus.* i. 123 a.
[3] *C.R.R.* ii. 76.

of the first year of Henry I,[1] in which scutage is mentioned *eo nomine*, proves that it was already in use when that king came to the throne. Scutage was, of course, the commutation of military service at a fixed rate, usually a pound or 2 marks on the knight's fee. The tenant-in-chief was responsible for rendering the required sum at the Exchequer, but he recovered it from his own tenants. Thus he was saved the trouble and anxiety of providing a properly trained knight, and it cost him nothing. In fact he might make a profit out of the transaction. It happened in·this way: many barons, as we have said, had enfeoffed more knights on their estates than were required to carry out their *servitium debitum*. These could collect scutage from all the enfeoffed knights, and pocket the sum collected from the knights in excess of the *servitium debitum*. The service due from Bury St. Edmunds was forty knights; the abbot had enfeoffed fifty-two. The result was, as Jocelin of Brakelond explains, that 'so often as 20s. are charged upon the fee, there will remain £12 to the abbot; if more or less is assessed, more or less will remain as a balance to him'.[2] Scutage, therefore, might be not only an easy, but possibly a profitable means of escape for the tenant-in-chief from his obligations.

Towards the close of the twelfth century an alternative method of commutation was devised. The tenant-in-chief could proffer a fine, a sum fixed arbitrarily by the king, which would relieve him of the whole burden of his military service; and its acceptance by the king carried with it an authorization for the vassal to collect scutage at the current rate from his sub-tenants. The precise relation of the fine to scutage is a matter which still requires elucidation. We cannot, I think, be content with the most recent explanation that the fine was merely an alternative method of commutation.[3] It was usually greater, often

---

[1] Printed by W. A. Morris, *Eng. Hist. Rev.* xxxvi (1921),·45.

[2] *Chronica* (Camden Society), p. 49.

[3] Chew, op. cit., p. 49: 'The difference between the fine and the scutage was originally one of form rather than principle.' But an entry on the Memoranda Roll· of 1199 (Pipe Roll Society, N.S. xxi, p. 76) shows that they were regarded as distinct: 'There must be discussion with the justices about the scutage of William Fitz Alan. The sheriff shows the king's writ in which is contained that he fined for 60 marks for his knights, but he does not say for scutage, and those 60 marks he pays and wishes to be computed for scutage.' Some tenants paid both scutage and fine. For example, all the religious houses in Dorset and Somerset did so in 1206 (*P.R. 8 Jo.*, p. 134).

very much greater, than the scutage. Some substantial ad-
vantage must have been gained from paying this heavier sum.
Maitland held that, while scutage relieved the baron of recruiting and sending his knights into the army, it did not absolve
him personally from service. The fine relieved him of both
obligations.[1] There seem to be no valid grounds for rejecting
this view. But perhaps it only represents a part of the truth.
It is unlikely to be a mere coincidence that fines first appear
on the Pipe Rolls in 1195, just about the time when the barons
were beginning to protest against service overseas. We are all
familiar with the emphatic words uttered by St. Hugh of
Lincoln at the council held at Oxford in 1197: 'I know', he
said, 'that the church of Lincoln must perform military service
to the lord king, but in this land only; outside the boundaries
of England no such service is due.'[2] Historically the bishop's
assertion was ill founded; but there is no doubt that he was
expressing an opinion widely held among the barons. There
is evident reluctance on their part to spend their energies in
long and indecisive wars on the Continent. These fines are
variously described on the rolls as fines for 'not crossing in the
army of Normandy' (*ne transfretet in exercitu Normanniae*), or 'for
having peace when he does not cross' (*pro habenda pace ne transfretet*), or 'for leave to remain in England' (*pro licentia remanendi
in Anglia*). Later they are said to be for (or instead of) his
passage (*pro passagio*). All these phrases point unmistakably to
service overseas. It is only at a further stage that we hear of
fines *pro servitio*, and the principle is then applied to service in
Wales and Scotland. The barons did not show the same reluctance to personal service at home, and in fact in the campaigns of John against Wales and Scotland the feudal levy
turned out in considerable force. Relatively only a few offered
fines to be relieved of their obligations.

Commutation by fine seems to have been in origin a compromise. The barons, on their side, were reluctant to serve
abroad; the king, on his, was prepared to concede this point.
The more readily perhaps because he no longer wanted so
many knights as in the wars of the past. The knight had ceased
to form the rank and file of the army. The king required only

---

[1] Pollock and Maitland, *Hist. of Eng. Law*, i. 269.
[2] *Magna Vita S. Hugonis*, ed. Dimock, p. 249.

a few of them to act as officers in command of hired men-at-arms, but who were prepared to remain in the army for a long term of service. Already in 1197 and again in 1205 only a fraction, a quota of the service due, was summoned to the host; and this quota gradually superseded altogether the ancient *servitium debitum* as far as actual service in the field was concerned. But if the barons and their knights were to get out of service abroad, they must make a contribution, and a substantial contribution, to the cost of the war in return for their exemption. The construction and upkeep of fortifications, and the maintenance of a large (or large for those days) army of hired knights and men-at-arms, were expensive. It is not unnatural that the king exploited the desire of the barons to rid themselves of their military obligations by driving a hard bargain.

These fines conform to no rule. The abbot of St. Augustine's, Canterbury, fined with £40 for his fifteen fees in 1197, when the scutage rate was a pound on the knight's fee. He could recover, therefore, the scutage of £15, which leaves a balance of £25, which presumably represents the commutation of his personal attendance and his contribution to the expenses of the war. Hugh of Bayeux also fined for £40, but on twenty fees, which, after the deduction of the scutage, leaves a balance of £20.[1] Hugh, bishop of Coventry, who had been disgraced for the part he had played in the rebellion of John in 1194, and who therefore had no claim on the king's generosity, fined in 1196 with £25 'that he need not cross with the army and for having the scutage of 25 knights'. He would, therefore, recover the whole of his fine from his under-tenants and pay nothing himself.[2] The arbitrary character of these fines defies explanation. We shall, perhaps, better realize why the barons were prepared to submit to such exactions if we examine the position of a tenant when he received a writ of summons to the host and consider the difficulties which faced him.

The service due from the small Somersetshire abbey of Muchelney was a single knight. In his return of 1166 the abbot reports to the king that he has enfeoffed two persons for the performance of this service of one knight: Richard Revel a part,

---

[1] *P.R. 9 Ric. I*, pp. 30, 111.
[2] *Chancellor's Roll, 8 Ric. I* (P.R. Soc. N.S., vol. vii), p. 59.

and Margaret, the daughter of Ralph Tabuel, a part.[1] For the
organization of their joint service, it must have been a serious
inconvenience that their estates lay in different parts of the
county, at least thirty miles apart, the one at Downhead on
the eastern side near the Wiltshire border, the other at Ilminster
to the south near the border of Dorset. From the record of
a case which came before the king's court nearly fifty years
later in 1210, we know how the duty was shared between them.[2]
The fee at this date was still held by the same Richard Revel,
who must have been well over seventy, and another lady,
Christina of Wick, who was clearly the principal tenant, for she
was required to find 'the body of one knight and the arms of
one knight and three-fifths of the maintenance (*estoverium*) of one
knight'. The remaining two-fifths of the maintenance was the
full military obligation of Richard Revel. An old man and a
woman were not the best-suited pair to be responsible for the
military service due from this little abbey. It is not surprising,
therefore, that the abbot generally commuted his service by
scutage or fine. During the first seven years of the reign of
John he paid scutage on three occasions at the current rate
of 2 marks; twice he fined with 10 marks and once with 5 in
addition to an *auxilium* of 10 marks; and in 1206 he fined with
£2 and paid the scutage in addition. But for the Scottish campaign of 1209, like the majority of the tenants-in-chief, he relied
on his sub-tenants to provide the service—with disastrous
results. On the Pipe Roll of this year it is recorded that the
abbot of Muchelney is amerced 100 marks because he neither
came nor sent his service.[3] He accordingly took proceedings
against his defaulting tenantry. Richard Revel essoined or
excused himself from attendance at court on the ground of
sickness (the recognized essoin *de malo lecti*);[4] but before he died
three years later he had made the most binding acknowledgement of his obligation in a final concord in which he admitted
that 'he owed the abbot and his successors two parts of the
service of a knight in money when five parts go to make the
service of a knight; . . . and be it known that if Richard or his
heirs do not do the said service, the abbot or his successors may

---

[1] *Red Book of the Exchequer*, p. 224.   [2] *C.R.R.* vi. 79.
[3] Pipe Roll of 1209, quoted Mitchell, *Studies in Taxation under John and Henry III*,
p. 96.   [4] *C.R.R.* vi. 75.

distrain upon the said tenement in Downhead'.[1] Christina, the other tenant, duly appeared in court to answer why she had not found a knight and arms and three parts of the maintenance of the same knight as she ought to have found. She admitted her guilt, and was amerced half a mark in addition to making good the loss incurred by the abbot.[2]

This case of the abbot of Muchelney, besides illustrating the difficulties involved in the system of knight service, shows also that the king was aware of the embarrassment in which his tenants-in-chief might be placed and was prepared to help them by enforcing in his courts the private bargains which they had made with their sub-tenants. But it was a troublesome and costly business. Here the king lost the service of a knight in his campaign to Scotland; the abbot was involved in tedious litigation; and his lady tenant had to pay a fine and damages. In short, the whole system was cumbersome, unreliable, and difficult to work. It could not be abandoned, because the whole structure of feudal society was based on it; and, as is well known, it survived until the reign of Charles II. But it was in the power of the king to take money instead of service; and this was very often preferable to all parties concerned, the king, the tenant-in-chief, and the sub-tenant. It obviated all the difficulties.

The exaction of military service was rendered enormously more complicated as the twelfth century progressed by the splitting up of fees by subinfeudation. If we read the records concerned with the upper- and middle-class society of this period, it is comparatively rare that we meet with tenants of an integral knight's fee. The fees have been broken up into bewildering particles. The description of a parcel of land as 'a fifth part of a third part of a fee of one knight'[3] illustrates how estates had ramified in the course of years. It is easy to imagine that on a well-organized estate a group of tenants holding large fractional fees, a half, a third, or a quarter, might make some arrangement among themselves for the performance of the service by a plan of election or rotation, a practice which is known to have obtained at the abbeys of St. Albans and

---

[1] *Feet of Fines* (Somerset Record Society, vol. vi), p. 27; *Muchelney Cartulary* (ibid., vol. xiv), p. 97.

[2] *C.R.R.* vi. 79.   [3] *C.R.R.* vii. 156.

Ramsey and elsewhere. But it is scarcely conceivable that such a system could be established for holders of minute and scattered fractions. The service of these could only be reckoned in terms of cash. That this was done at an early date is evident from a charter preserved among the manuscripts of Sir Christopher Hatton.[1] As this charter was granted by Peter who was the son of a Domesday tenant, William son of Corbucion,[2] it can hardly be later than the reign of Henry I. It gives land at Wolverton in Warwickshire to William son of Reinfred and his heirs to be held 'by a third part of the service of one knight in such a way that he shall acquit his whole service by the yearly payment of twenty shillings'.[3] A precise statement like this at so early a date must be very rare, if not unique. But by the end of the twelfth century, when the process of multiplying fractions had proceeded far, there is more evidence to the same effect. In a final concord made between Robert Blundus and Roger of Ginges, which is enrolled on the Pipe Roll of 1197,[4] it is stated that Robert and his heirs have and hold certain lands by 'the service of a tenth part of a fee of one knight and a pair of gilt spurs or sixpence rendered within the octave of Easter . . . and they shall do the aforesaid service of a tenth part of a fee of one knight by money (*per denarios*), whenever the same Robert and his heirs ought to do that service'.

A very clear illustration of the system at work is provided by a record of the contributions of the tenants of the barony of Lovetot to the scutage of Gascony assessed in 1242–3 at 3 marks on the fee. Three gentlemen in Huntingdonshire held among them two-thirds of a knight's fee. The three holdings are in different villages, fairly widely scattered, and each is carefully described as 'a third part of two parts of the same knight'. Each of the three tenants contributes 8s. 11d., which, added together, makes £1. 6s. 9d., one penny more than 2 marks, the correct amount of scutage from two-thirds of a fee.[5]

That scutage from the point of view of the small under-tenants was nothing more than an additional tax added on to

---

[1] *Sir Christopher Hatton's Book of Seals*, Hatton MS. no. 528. I am indebted to Professor Stenton, who not only supplied me with this reference but also with a transcript of the charter.

[2] *D.B.* i. 243 b.

[3] Cf. also Dugdale, *The Antiquities of Warwickshire*, p. 467.

[4] *P.R. 9 Ric. I*, pp. 167–8.   [5] *Book of Fees*, p. 928.

the rent is thrown into the clearest light by an agreement made
between Thomas Fitz William and Roger of St. Philibert concerning land at Brampton in Suffolk, which is entered on the
Curia Regis Roll in 1206.[1] It sets out the names of twelve
sub-tenants (of whom Roger of St. Philibert seems to be the
principal one) with their respective rents and the proportion
of scutage that they are required to contribute. Thus the first
tenant, Walram of Stofne, 'owes 11s. 6d. of rent, and to twenty
shillings of scutage he owes 25d., to more, more, to less, less;
and at Christmas four hens and one goose'. The scutage contribution is not, however, determined in a fixed proportion to
the rent. Robert Pulcre, another tenant, pays only 6d. and
a couple of fowls in rent, but he must contribute 6d. also to
the scutage.

Robert Pulcre must have been a man in a very humble position; and his contribution to scutage strikingly demonstrates
the fact that this money commutation for fighting ultimately
rested, not on the fighting class, the aristocratic class of knights
who held their lands in return for performing military service,
but on men near the lower margin of society. How deeply this
burden might penetrate appears from a complaint brought by
the villagers of Wicken in Northamptonshire before Archbishop
Hubert Walter and his fellow judges in 1196.[2] Their complaint
was that the manorial lord Peter Fitz Ralph and Alice his wife
were in the habit of exacting scutage from them. They appeared
in court and acknowledged that they were villeins and customary tenants of the said Alice by performing whatever works
she might order and by giving merchet for the marriage of
their daughters (the recognized tests of villeinage); and they
go on to declare that of their own accord (*sponte sua*) they
entered into an agreement to pay scutage on this occasion at
20s. and at other times as they might be able to arrange. The
case of the Wicken peasants was, we may hope, exceptional.
Villeins were liable to be tallaged at the will of their lord. If,
in addition to this, a contribution to scutage was generally
exacted, their financial burdens would be almost intolerable.

Although a contribution to scutage was the normal obligation of those tenants who held fractions of knight's fees, sometimes small but irritating intrinsic services might be also

[1] *C.R.R.* iv. 134. [2] *C.R.R.* i. 16.

demanded. Occasionally they were the subject of litigation. In 1203, for example, two of the tenants of Henry Tresgoz in Sussex, William Mordant and William of Billingshurst, put themselves on the grand assize. The former sought recognition whether he ought to render 18*d.* of scutage (when scutage was at 20*s.* on the fee) for his whole service, as he acknowledged, or whether besides the said money he should provide his lord with a dinner every year and a horse in the king's army when he demanded it, as Henry exacted. William of Billingshurst questioned whether, in addition to scutage on the tenth part of a knight's fee, he ought to be required to provide an annual dinner and a horse to take his lord's small child (*infantem*) from one manor to another.[1] These services were perhaps trivial, but if they had not been felt as something of a hardship, these tenants would scarcely have gone to the trouble and expense of contesting them in court. The normal issue of a grand assize was a licence of the court for the parties to settle the matter between themselves. So it was in these cases of Henry Tresgoz and his tenants. The dinner and the horse for the king's army demanded from William Mordant were commuted for a pound of pepper; the dinner and the child's horse required of William of Billingshurst were discharged by the annual render of a pair of gilt spurs or 6*d.*[2] The substitution of these small renders in place of the actual performance of definite services is a further sign of the breakdown about the turn of the century of the early feudal arrangements.

In the early days of feudalism possibly the most onerous of the duties which were often attached to knight service was that of providing the garrison of the royal or baronial castles, the service of castle-guard. In a famous passage Jocelin of Brakelond has described the organization of the guard duty required from the knights of the abbey of Bury St. Edmunds.[3] In 1196 the forty knights were ranged in four groups, called constabularies, of ten knights each, to mount guard at Norwich castle. The substance of Jocelin's statement is correct, for it is corroborated by an official return, an inquest of service belonging

---

[1] *C.R.R.* ii. 142–3, 253; iii. 5. For a somewhat similar additional service of much earlier date see Stenton, *English Feudalism*, p. 188.

[2] For the final concords which record these settlements see *Sussex Fines*, ed. Salzman (Sussex Record Society, ii), nos. 79, 89.

[3] p. 49 (Camden Society).

to the early years of the thirteenth century, preserved in a manuscript in the Harleian collection.[1] This list gives not only the names of the knights in each constabulary almost identically as in Jocelin's list, but also designates one of the ten as the commander, the *constabularius*. In fact, however, this organization of the Bury knights is somewhat illusory. No doubt at one time they actually went on sentry-go in their squads of ten. But at the time that Jocelin was writing it is already a thing of the past. The service is already commuted, and the knights are haggling with the abbot as to whether they ought to contribute 2s. 5d. or 3s. A surviving return of the justices in eyre, who visited Norfolk two years later (1198), shows three of these Bury knights owing these small sums for the ward of Norwich castle.[2] In fact the castle was garrisoned by a paid force, which in 1193 was composed of 25 knights, 25 mounted, and 25 foot sergeants.[3] It was to the wages of this little garrison, amounting to 37s. 6d. a day, that the knights of St. Edmund, and a number of others besides, were required to contribute.

It had indeed long since become the general practice to commute the service of castle-guard for money. Like scutage, it often came to be little else than an addition to rent, the payment of which continued for centuries after the castles themselves had crumbled into ruins, or at least ceased to be an essential feature of the defensive system of the country. As late as the reign of Queen Anne, for example, the manors of Tubney and Wytham, fields at Pusey and Sugworth, the village of Uffington, a dairy farm at Bagshot, and many other places scattered about Berkshire were still paying a pound or a few shillings 'ward money', representing the ancient obligation of the tenants of the abbey of Abingdon to furnish a guard at Windsor castle.[4]

A special significance has been attached by Round to the passage in Jocelin's chronicle to which I have referred. He thought that in the constabularies of the Bury knights he had

---

[1] Brit. Mus. Harleian MS. 645, fols. 25 and 222. It is printed in part in *Eng. Hist. Rev.* xliv (1929), 371 by Professor Galbraith, who has kindly lent me a transcript of the whole.

[2] *Book of Fees*, pp. 1326–8.   [3] *P.R. 5 Ric. I*, p. 13.

[4] *Journal of the House of Commons*, xiii. 204 (1700), quoted Round, *Archaeological Journal*, lix. 157.

identified the unit of the feudal host. 'I maintain', he emphatically writes,[1] 'that the extent of the obligation (of knight service) was fixed in relation to, and expressed in terms of, the *constabularia* of ten knights, the unit of the feudal host.' Round's argument here is hardly convincing; he has little beyond the Bury evidence on which to base it. The word constable was applied in the Middle Ages to almost any person in a position to command men, from the Constable of England, one of the highest household officers, to the village constable of thirteenth-century peace ordinances. It was used for a ship's captain and, as we shall see, for an officer in command of a large company of soldiers in the field. Most commonly it denotes an officer in command of a castle, the garrison of which, varying according to its size and importance, might be described as a *constabularia*. In the ordinance drawn up for the defence of the kingdom in 1205[2] the chief constables of the county (*capitales constabularii comitatus*) are directed to have the names in writing of the special constables (*particulares constabularii*) of the hundreds, boroughs, vills, and castles, and 'the number of armed men of each constabulary'. From these examples it is evident that the words *constabularius* and *constabularia* were indiscriminately applied to any commander and the body of men under him. There is no valid reason, therefore, to identify the *constabularia* of the knights of St. Edmunds with the unit of the feudal host.

Nevertheless, there is abundant evidence to prove that the knight was associated with a unit in the field which is likewise known as a *constabularia*. In the Pipe Rolls there are scores of entries recording fines paid into the Exchequer 'because so and so was not in his constabulary'. Thus in 1198 Walter of Cormeill renders account of 50 marks 'because he was not found in his constabulary', or William Painel in the same year accounts for £80 'for having his land of which he was deprived because he was not found in his constabulary in Normandy'.[3] On the roll of 1230 there is a note of several loans (*praestita*) made in the time of King John. Among them is an entry, 'William Marshal, earl of Pembroke, for making advances to the knights

---

[1] *Feudal England*, p. 261. The statement is repeated by Round in his article on Castle Guard (*Archaeological Journal*. lix (1902)) and in *The King's Sergeants and Officers of State* (1911), p. 76.

[2] Gervase of Canterbury, *Gesta Regum*, ed. Stubbs, ii. 97.

[3] *P.R. 10 Ric. I*, pp. 214, 195. Cf. also ibid. 16, 42.

of his constabulary'.[1] In 1193 the same earl accounts for receipts and expenses incurred in the time of war for carrying out the king's business. Among the items are payments for a force of 10 knights and 500 foot at Bristol, for another of 20 knights, 40 mounted archers, and 450 infantry at Gloucester, and yet another body of 500 foot at Windsor.[2] Here are three companies of an approximate strength of 500 men. In 1195 there is a charge on the Hampshire account for the transport of 1,300 foot and 3 knights as 'masters' (*magistri*).[3] In the roll of the following year the sum of 100 marks is allowed to William de Marisco, a well-known captain in the Norman wars, 'for paying 500 Welsh infantry of his constabulary', and a further sum of 3 marks to William himself 'their constable'.[4] Another payment is made for 500 Welsh foot and 5 horse and 'to one knight, their constable'.[5] A third entry relates to another body of 500 Welsh foot-soldiers crossing into the king's service, 'and to Philip, the knight of Robert de Boilliers, their master'.[6] From these notices we are justified in assuming that the knights were commonly employed as officers in command of bodies of troops composed chiefly of infantry, but sometimes including a few cavalry; that they were called indiscriminately constable or master (though the latter term is also used for what we should describe as non-commissioned officers); and finally, that the unit of their command, the *constabularia*, was approximately 500 men.

A lord might stipulate that one of his knights should undertake to serve as constable. After a detailed description of the fee of Richard Cotele in the Glastonbury inquest of 1189, it is added, 'All this he holds by the service of one knight, and if the lord wills he will be constable (*constabulus*)'.[7] There were also tenants on the Welsh border who held their lands by sergeanty for the service of being *constabularii peditum*. A Shropshire tenant, for instance, Roger de Lamere, owed the service of being constable of infantry in the king's army in Wales at the king's cost at 12*d*. a day—the normal rate of pay for a knight at the end of the twelfth century.[8]

[1] *P.R. 14 Hen. III*, p. 129.
[2] *P.R. 5 Ric. I*, p. 148.
[3] *P.R. 7 Ric. I*, p. 205.
[4] *Chancellor's Roll, 8 Ric. I*, p. 19.
[5] Ibid., p. 88.
[6] Ibid., p. 42.
[7] *Liber Henrici de Soliaco* (Roxburghe Club), p. 5.
[8] *Book of Fees*, p. 145; cf. also pp. 347, 1183, *constabularius ducentorum hominum peditum*.

Commutation of service on a large scale, as we find it at the turn of the century—some 4,000 knights secured exemption from service in 1205 by scutage or fine—meant that the feudal levy had ceased to be an effective fighting force. It no longer met the requirements of war; it was already out of date. It was superseded by an army chiefly composed of men paid to fight. This very significant change in military organization is reflected in our word 'soldier'. The word is derived from the medieval Latin word *solidarius*, a man who serves for pay. The English noun 'sold', now obsolete, but common down to the sixteenth century, means pay; the Latin word for a shilling is *solidus*. The receipt of the king's shilling, or a day's pay, is even now the final and irrevocable act which makes a civilian into a soldier. Though we cannot press an exact connexion between the soldier and the shilling wage, the shilling was the normal day's pay of a knight at the end of the twelfth century. But like all wages it tended to rise. When in the Pipe Roll of 1162[1] we have the earliest mention of *milites solidarii*, their pay was 8*d.* a day; and in the reign of John the usual rate was not 1*s.* but 2*s.* These were high wages for those days, and necessarily so. The knight had to provide his own equipment. His charger (*dextrarius*) was an expensive beast; it could not be bought for less than 10 marks; and his armour of chain-mail, by reason of its elaborate workmanship, cannot have been cheap. The knight belongs to a limited class of men of substantial wealth. As we have seen, the records which throw light on the composition of armies at the end of the century suggest that, apart from the knightly retinues of the king or of a great baron like William Marshal, the knights were employed, not as fighting in divisions by themselves, but as officers in command of considerable bodies of men-at-arms. Their pay is that of colonels not of privates. These ordinary soldiers, mounted or foot sergeants, as they were designated (*servientes equites* and *servientes pedites*), could of course be recruited at much cheaper rates than the knights; the cavalryman received 4*d.* or 6*d.*, and the infantryman 2*d.* or 3*d.* a day. They belonged to a humbler class of society, and their equipment was correspondingly less elaborate than the demands of chivalry would allow. Essentially the soldier is not a man who serves in the army by reason

---

[1] *P.R. 8 Hen. II*, p. 53.

of his tenure of land, but one who is paid for fighting, a mercenary. And it is of such that the armies of this period were chiefly composed.

The knights who compounded for their military obligations found themselves fully occupied in the administrative work of their counties. The judicial business was largely thrown upon the knights; the grand assize, which was becoming increasingly popular, was always in their hands. Four knights acted as a committee to elect twelve other knights to serve on the jury.[1] The number of knights in any county was not large. In the early years of King John's reign there appear to have been about a hundred knights in Northamptonshire available for this duty.[2] In consequence the same men had to act repeatedly; the same names recur in case after case; several served on seven or eight juries, and one of them, Robert of Holdenby, on eleven in the five years between 1200 and 1204.[3] The grand assize was always a long and tedious action, subject to continual postponements for one cause or another, before it was ended, usually by an agreement between the parties in a final concord. Each case, therefore, would probably involve the jurors in a number of attendances. The case of one Northamptonshire defendant, who put himself on the grand assize in 1201, was ten times before the court before it was terminated in 1205.[4] It is not surprising that often the jurors 'made many defaults' (*fecerunt plures defaltas*).[5]

Normally ordinary freemen, *legales homines*, were regarded as competent to serve on juries to take the petty assizes. Sometimes, however, knights were required for such cases also. Thus an assize of novel disseisin was respited in 1204 'because the

---

[1] The electing committee who 'chose other than knights' were in mercy (*C.R.R.* v. 109). Sometimes they chose themselves (ibid. ii. 252–3).

[2] This figure is arrived at from an analysis of the grand assize juries between 1200 and 1204, years, of course, of continuous war on the Continent.

[3] *C.R.R.* i. 288, 326, 341; ii. 27, 157, 252, 253; iii. 195; Northamptonshire Assize Rolls, nos. 463, 467, 598.

[4] This was a case between Henry de Alneto of Cornwall and Henry de Alneto of Maidford concerning a knight's fee with pertinences in Maidford (Northants). The case opened in Trinity Term 1200 (*C.R.R.* i. 205); the defendant put himself on the grand assize in the Easter Term 1201 (ibid. 473); the jury was elected the next Michaelmas (ibid. ii. 27). Henceforward the case was continually coming before the courts until Easter Term 1205 when it was terminated in favour of the sitting tenant, Henry de Alneto of Maidford (ibid. iii. 341).

[5] Ibid. iii. 255; iv. 21.

king orders that the assize be taken by knights';[1] another case was postponed because the recognitors 'were not all knights'.[2] When the king was an interested party, the jury would naturally be formed of knights. So, when the king claimed the advowson of the chapel in the royal castle of Hereford against the prior of Hereford, 'all the recognitors were removed, and it was ordered that knights shall replace them'.[3] Evidently jury service was a heavy burden on both knights and freemen. In the thirteenth century favoured subjects obtained charters granting exemption from the obligation to serve,[4] with the inevitable result that the burden on the rest became still heavier. By 1258 the position had become so serious that it was said that in several counties the grand assize could not be taken owing to lack of knights.[5]

Serving on juries was only one of many administrative duties imposed upon knights. If a litigant excused himself from attendance at court on the ground that he was confined to his bed by sickness (the essoin *de malo lecti*), four knights must visit him and afterwards testify in court to the genuineness of his complaint;[6] they had much work to do in the appointment of attorneys. In pleas relating to land they were often called upon to inspect the ground—the tenement, wood, marsh, or whatever it might be which was in dispute. They must perambulate the boundaries and they must do it thoroughly, *de loco ad locum, de domo in domum*.[7] Their services were demanded to make assignments of dower and marriage portions,[8] to make valuations of land in a settlement by a final concord;[9] or they might be required to audit the accounts of an estate in wardship.[10] If

---

[1] *C.R.R.* iii. 224. [2] Ibid. i. 415–16.

[3] Ibid. vi. 28. Cf. vi. 234; viii. 364–5.

[4] In 1219 another knight had to be found in place of Alan of Wilton, the fourth knight 'qui amovetur pro carta domini regis Johannis que eum aquitat de omnibus assisis' (*C.R.R.* viii. 99).

[5] Petition of the Barons, c. 28. The situation was partially remedied in the Provisions of Westminster, c. 8.

[6] Though Glanvill (i. 19) says that two knights were sufficient for this purpose, the judges seem to have held that all four must view the infirmity. In a Somerset case (*C.R.R.* iv. 21) when three knights only made the view, the fourth was in mercy. When a sheriff sent *pauperes non milites*, he was reprimanded (ibid. i. 203).

[7] Ibid. vii. 194. [8] Ibid. ii. 63–4, 77; vii. 312.

[9] Ibid. iv. 237.

[10] 'Vicecomes faciat compotum audiri per legales milites de comitatu' (ibid. i. 407).

a case was transferred from one court to another, from the hundred to the county, from the county to Westminster, it fell to appointed knights to bear the record of the proceedings in the lower court, a duty involving much time and expense in travelling.[1]

In criminal cases their work was scarcely less arduous. Some would be required to act as coroners, and their time would be wholly occupied with the manifold duties connected with this office.[2] Heavy police duties were imposed on the knights 'assigned' to take the oath of keeping the peace under Hubert Walter's ordinance of 1195, and these duties were increased by subsequent ordinances during the thirteenth century. All offenders were to be delivered to these knights, who in their turn were responsible for delivering them to the sheriff.[3] Sometimes the sheriff might commit an offender to the custody of other knights.[4] Knights were also sent to inspect the scene of a crime[5] or to view the wounds of a victim of assault and measure them and afterwards give evidence in court.[6] If a case came to a trial by battle, it was knights who kept the ring and acted as seconds;[7] they must also make a record of the combat.[8]

Yet the authority of these knights, upon whom so much responsibility was thrust, was very restricted. A group of

[1] For a very elaborate record of a Cornish case transferred from the shire to Westminster see *C.R.R.* vii. 169 ff. This was carried by two knights who were present when the record was made.

[2] Cf. the articles of the eyre of 1194, c. 20: 'in quolibet comitatu eligantur tres milites et unus clericus custodes placitorum coronae' (Hoveden, iii. 264). By 1198 all four are spoken of as knights; an appellor shows his wound to William de Baus 'uni iiii' militum custodum placitorum corone' (*C.R.R.* i. 39). Two coroners in Northumberland held their office by sergeanty (*Book of Fees*, p. 1149). Cf. E. G. Kimball, *Sergeanty Tenure in Medieval England*, pp. 84 ff. But these were exceptional.

[3] Hoveden, iii. 299–300. Cf. C. A. Beard, *The Office of Justice of the Peace in England in its Origin and Development*, pp. 17 ff.

[4] *C.R.R.* vii. 95. The sheriff commits an accused man to the custody of six lawful knights.

[5] In a case of housebreaking, for example, 'milites missi ex parte Justiciarum ad videndum si delictum factum esset, . . .' (*Northamptonshire Assize Rolls*, no. 492).

[6] Cf. *C.R.R.* i. 39; *Pleas of the Crown* (Selden Society, vol. i, ed. Maitland), no. 64. It was always insisted upon that wounds should be shown when fresh. The measuring of wounds was generally left to the coroners. Bracton stresses the importance of precise measurements of length and depth, for if it is only a slight wound there can be no appeal (ff. 144, 145).

[7] *C.R.R.* i. 100. See Hans Fehr, *Das Recht im Bilde*, plates 40 and 52, where illustrations of duels showing the seconds in the ring are reproduced from a Bern MS.

[8] *C.R.R.* vi. 67.

knights, for example, by a skilful piece of detective work traced certain robbers by following the hoof-marks of the horses on which they tried to escape (one of the horses was unshod on the hind feet); but when they came up with one of the criminals 'they dared not do anything since they had not with them the king's sergeant', who arrived on the scene too late.[1] On the other hand, if they failed in the performance of their duties, they were liable for punishment. They were fined for wrong pleading (*pro stultiloquio*)[2] or if they made a bad record of a case before a higher court.[3] If as jurors they gave a wrong verdict, had sworn falsely in an assize, even in the grand assize (despite the contrary view of Bracton), they could be attainted by a jury of twenty-four, and be punished by the loss of their chattels and at least a year's imprisonment.[4]

Nevertheless it is certain that many of the knights, who laid aside their arms and· armour and took to the life of country gentlemen,[5] like the justices of the peace in the succeeding age, were willing enough to shoulder these burdens, enjoying the prestige and influence which they gave them. There were always men who liked the exercise of authority and the opportunity to intrigue, who busied themselves with judicial matters (*buzones*).[6] Already by the end of the twelfth century knights were training themselves in the arts of local government which in the following centuries they were so largely to control.

---

[1] *C.R.R.* vi. 23-4.   [2] Ibid. vii. 195.

[3] 'Et quoniam timuerunt milites male recordatum fuisse, dederunt i marcam' (ibid. v. 17).

[4] Glanvill, ii. 19. Bracton (f. 290) maintains that jurors of the grand assize cannot be attainted. But an Oxfordshire case taken in 1206 (*C.R.R.* iv. 118, 141, 173) proves the correctness of Glanvill's statement. Cf. G. E. Woodbine's note on the passage in his edition of Glanvill (p. 204).

[5] Richard Revel, a Somersetshire knight, repeatedly emphasized in court his *gentilitas* and that he and his father and his brothers were *naturales homines et gentiles de patria* (*C.R.R.* iii. 129).

[6] Ibid. vi. 231; Bracton, f. 115. I follow Mr. C. T. Flower's rendering of the passage *qui sunt buzones judiciorum*: 'those making a business of justice' (*C.R.R.* vi. 525). Cf. Pollock and Maitland, *Hist. of Eng. Law*, i. 553. For an elaborate discussion of the word *buzones*, but with a different interpretation, see G. T. Lapsley, *Eng. Hist. Rev.* xvii (1932), 177, 545.

IV

SERGEANTS

IN the course of these lectures I have often had occasion to refer to the heterogeneous group of tenants who held their estates by what is termed sergeanty. It is so heterogeneous that it defies any clear classification. Some sergeants are barely distinguishable from knights, others from tenants in socage, and others again from those who held by the spiritual service of frankalmoign. An early example of the close identity of sergeanty and knight service comes from the Black Book of Peterborough, a survey of the estates of the abbey made between 1125 and 1128.[1] It is there recorded that Abbot Thorold, who died in 1098, gave to a certain Vivian a sixth part of a hide in Oundle and a quarter of a hide in Warmington in sergeanty (*in sergentaria*), for which he ought to be knight in the army with two horses and his own arms, and the abbot will find for him other necessaries. He holds by sergeanty, and yet he must perform for his sergeanty the normal service of a knight. A curious example of the confusion of tenures is afforded by the tenant of Papworth St. Agnes in Cambridgeshire. He was required for his service to feed two paupers daily for the safety of the souls of the king and his ancestors and successors. Was this sergeanty or frankalmoign? In a charter of 1208 it is entered as a sergeanty;[2] on the Fine Roll of the same year it is described as frankalmoign.[3] A Cambridgeshire jury, to which the question was referred, decided on frankalmoign.[4] And yet in the records of the middle of the thirteenth century it is still described as a sergeanty.[5] On the other hand, there was no doubt as to the form of tenure of a group of tenants who held half a hide of the king at Apps in Walton-on-Thames. Their service was similar, but rather more serviential in character, that of making and distributing in charity a cuve (*cuvata*) or

---

[1] Printed as an Appendix to the *Chronicon Petroburgense* (Camden Society), p. 175. Cf. *Henry of Pytchley's Book of Fees* (Northamptonshire Record Society), p. 120.

[2] *Rot. Chart.* 180 a.

[3] *Rot. de Obl. et Fin.*, p. 425.

[4] *C.R.R.* v. 200.

[5] *Book of Fees*, pp. 1181, 1234.

cask of beer on All Saints Day. It is always entered as a tenure in free alms.[1]

Between sergeanty and socage the dividing line was so slender that the barons of the Exchequer themselves were at a loss to know how to draw it. Referring to the estate of Robert Bussel in Devonshire they write to the king:

'Having inspected at the king's command his charter to Theobald de Engleychevill of the manor of Teyngwyk, the said Theobald's charter to Robert Bussel, the king's confirmation thereof, and the eschaetor's inquisition, they find that the said Robert held the manor of the king by service of 1 pair of gilt spurs yearly for all service. They do not find from the rolls that land so held is socage, but land held by 1 pair of gilt spurs or 6d. or more is socage; it is for the king and council to judge whether land held by 1 pair of gilt spurs for all service is socage or sergeanty.'[2]

Even a money rent, the most common socage tenure, is some-times regarded as a sergeanty. A striking instance of this is provided by the case of Roger Grossus or Crassus, as he is variously called, Roger the Fat, who held an eighth part of a knight's fee in chief of the king in the village of Hacconby in Lincolnshire. He seems to have been a reckless and charitable person, and he dissipated his property in gifts chiefly to churches and monasteries. Some he gave to the monks of Sempringham, some to the Templars, some to his neighbours; and when he died in 1202, all that was left of his eighth part of a knight's fee was a toft and 1½ acres in a wasted condition worth 12d. a year.[3] Most, if not all, the forinsec service, the knight service, had been thrown on the recipients of his bounty, but what of the miserable wasted remnant? We should expect, if it were no longer knight service, that it would be socage. It is recorded, however, on the roll of the itinerant justices who visited Lincoln in 1202 that the sheriff, Gerard de Camville, must account for the land of Roger Grossus, which he held by sergeanty (per serianteriam) of the king by rendering 12d. a year.[4]

Sergeanty, then, was a hotchpotch of tenures that did not,

[1] Book of Fees, pp. 67, 273; later, additional services were added, ibid., pp. 1363, 1378.

[2] Cal. of Inquisitions p.m., Henry III, no. 714; E. G. Kimball, Sergeanty Tenure (1936), p. 116.

[3] The Earliest Lincoln Assize Rolls, ed. D. M. Stenton (Lincs. Rec. Soc.), no. 759. Cf. also nos. 479, 480, 482, 483, 485, 486; Book of Fees, p. 181.

[4] Linc. Assize Rolls, no. 1047 (p. 179); P.R. 4 Jo., p. 235.

for one reason or another, conveniently fit into the general
scheme of social grouping. But how in fact were they to be
distinguished from the other tenures to which they were so
closely akin? There were indeed some rules: a sergeanty should
not be alienated; it should not be divided among heiresses, for
normally it involved some specific personal service which only
an individual could properly perform. But by the thirteenth
century these rules were being disregarded. Round insists that
the differentiating mark between knight service and sergeanty
was the payment of scutage. 'The touchstone', he writes in his
monograph on the *King's Sergeants*,[1] 'by which in practice the
two tenures were distinguished was the payment of scutage',
and he continues: 'If a tenant was liable to scutage, his tenure
could not be serjeanty, and if conversely, it was sergeanty, he
was not liable to scutage.' In support of these statements he
quotes Bracton or rather Bracton's interpreter, Maitland. But
both Bracton and Maitland are in fact discussing, not the
difference between knight service and sergeanty, but between
knight service and socage.[2] Scutage in any case is an unsafe
foundation on which to base class distinctions. Many people
other than knights were liable, or at least were compelled, to
pay scutage; and the government not only sanctioned but
assisted in enforcing such payments. Thus in 1230 the king
ordered the sheriff of Lincolnshire to help Hilary Trussebut to
distrain his knights and free tenants (*milites suos et libere tenentes*)
to render their scutage.[3] Round no doubt intended his words
to refer only to the liability of tenants-in-chief, but even so the
available evidence scarcely warrants such a positive assertion.

From the end of the twelfth century, as we have seen, there
were two alternative methods of commutation of service—
scutage and fine. Owing to the almost infinite variety in value
and size of sergeanty tenures, the fine was obviously the more
appropriate form of commutation. The scutage was fixed in
relation to the knight's fee; the fine, on the other hand, could
be adjusted to size and value. That sergeants commuted their
service by fine is unquestioned. Too much emphasis, however,

---

[1] *The King's Sergeants and Officers of State*, pp. 22–3; cf. also p. 3.
[2] Bracton, f. 37; Pollock and Maitland, *Hist. of Eng. Law*, i. 277. Bracton's
words here are not free from obscurity.
[3] *Memoranda Roll, 14 Hen. III*, p. 91.

can be placed on this distinction. Evidently the clerks in the Exchequer regarded the scutage and fine as in essence the same. They are listed under the same headings. The third scutage of King John, for example, levied in 1202, appears on the Pipe Roll under the heading 'Of fines and scutage of knights of the third scutage'[1] or simply 'Of the third scutage',[2] and in the lists that follow knights and sergeants are thrown together indiscriminately. The Exchequer clerks, therefore, treated the scutage and fine as the same for the purposes of accounting. The unreliability of the scutage test can be plainly seen from the entries on the rolls relating to the sums due for scutage or fine from a group of Buckinghamshire tenants in the early years of King John:

	3rd scutage 1202[3]	4th scutage 1203[4]	5th scutage 1204[5]	6th scutage 1205[6]
Aubrée de Jarpen-vill (Marshal of the Hawks)	10 marks *de scutagio pro serianteria*	10 marks *pro serianteria sua*	...	...
Isabel de Clinton (Larderer)	5 m. *pro feodo 1 militis*	10 m. *de serianteria sua*	10 m. *pro serianteria*	10 m. *de fine*
Robert Mauduit (Chamberlain of the Exchequer)[7]	3½ m. *de secundo scutagio* and 3½ *de hoc scutagio*	...	...	...
Herbert de Bolebec (1 knight's fee)	3 m. *de scutagio et pro passagio*	2 m. *pro serianteria sua*	6 m. *de feodo 1 militis*	5 m. *de fine suo*

From this evidence, and especially from the entry relating to Aubrée de Jarpenville, the Marshal of the Hawks, it would appear that at this time some tenants in sergeanty were being charged with scutage. There is, moreover, other incontrovertible evidence which supports this conclusion. On the Memoranda Roll of 1230[8] there is a mandate addressed to

[1] *P.R. 4 Jo.*, p. 127 (Wilts.) and passim.   [2] Ibid., p. 80 (Hants).
[3] Ibid., p. 28.   [4] *P.R. 5 Jo.*, pp. 96–7.
[5] *P.R. 6 Jo.*, p. 15.   [6] *P.R. 7 Jo.*, p. 74.
[7] 'Robertus Maudut, Camerarius, tenet Hamslap'; nescitur an in seriancia an in milicia.' *Book of Fees*, p. 20. He received a writ of quittance in 1206 which would suggest that his tenure was military (*P.R. 8 Jo.*, p. 42).
[8] Pipe Roll Society, N.S. vol. xi, p. 47; cf. pp. 63, 95.

the sheriff of Gloucestershire ordering him to respite 'the demand which he makes of Ralph Moyne for several scutages of fees which he holds of the lord king, which fees he holds by sergeanty and not by military service as he says'. Ralph Moyne held Shipton in Gloucestershire together with estates in several other counties by the service of buying for the king's kitchen (*emptor coquine domini regis*). A somewhat similar mandate was sent in the same year to the sheriff of Oxfordshire bidding him hold over the demand

'which he makes of Roger, usher of the Exchequer, for our last scutage for the tenement he holds in Aston (Rowant), which the same Roger says that he holds by sergeanty and not by military service, and then make it known to the barons of the Exchequer plainly and distinctly on what ground he exacts from him the said scutage for the said tenement'.[1]

We may infer from these mandates that the sheriffs had been taking, or attempting to take, scutage from sergeanty tenants; and these tenants had protested; and that from about this time (1230), the rule was being established that these tenants were not liable to scutage. We can, perhaps, go farther than this. It appears that henceforward they were likewise exempt from the fine. They were exempt, that is to say, from both forms of commutation of service. But already by this date the system, on which tenure by sergeanty rested, was outgrowing its usefulness.

Sergeanty reflects a time when land was plentiful and money was scarce. It was easier to pay for the various services which the king or his great barons required by a grant of land than by a cash payment. It was in this way that the king provided for his household and administrative staff, for the multitude of servants needed for his pastimes of hunting and falconry, and in part for his army. Some few can be traced back to Domesday Book where they appear as *servientes regis*; many more vaguely claim that they have held their lands in this way from the time of the Conquest.

When in the reign of Henry I we get our first glimpse into the organization of the royal household in the famous document known as the *Constitutio Domus Regis*,[2] it is already partially

---

[1] Ibid., p. 63.
[2] Printed in the *Red Book of the Exchequer*, iii. 807 ff.

established on a wage basis. The king's household was then
composed of a strange assortment of distinguished administrators and menial servants. There is the chancellor and his
chaplains; there are bakers and butlers, cooks and scullions,
marshals and ushers, hornblowers and kennelmen; there are
huntsmen of all kinds, including wolf-hunters and cat-hunters
(*catatores*); there are also archers. Some of these have fixed
stipends and receive allowances of cakes, wine, and candle-ends.
Most of the domestic staff feed in the king's house without other
emolument, for, we may assume, they were paid for their services by the lands which they held in sergeanty; but they
received wages for their assistants, usually at the rate of three-
halfpence a day. The outdoor staff was also in receipt of money
wages. All these royal servants, however, did not hold their
positions by virtue of tenure in sergeanty. Of the heads of
departments, the steward, the constable, and the chief chamberlain certainly did not; and it is an open question whether the
marshal and the butler held their offices by sergeanty or by
knight service.

No principle seems to have governed the size of tenements
held in sergeanty; nor did they bear any relation to the magni-
tude of the services required. One man might hold a valuable
manor for a trifling service, another a few acres in return for
arduous work. Henry II enfeoffed Boscher, his servant, with
the manor of Bericote in Warwickshire with a mill and pertinences worth £5 a year by the light service of keeping a white
hound with red ears and delivering it to the king at the end
of the year and receiving another puppy to rear.[1] The king's
larderer at York, on the other hand, whose duties were many
and laborious, held for his sergeanty but two carucates at
Bustardthorpe worth. 10s. a year.[2] It was an old sergeanty
traceable to the time of Henry I, for King Stephen, early in his
reign, granted to John his larderer of York and David his son
all his land *cum ministerio suo de lardario*, as he held it on the day
that Henry died.[3] An inquisition of Henry III's reign informs
us of the duties assigned to this larderer in the twelfth century.[4]

[1] *Book of Fees*, p. 1278.
[2] Ibid., p. 1202. He had other lands both in the city of York and in the
neighbourhood, but they do not appear to have been attached to the sergeanty.
[3] Farrer, *Early Yorkshire Charters*, i, no. 243.
[4] *Cal. Inq. Misc.* i, no. 501.

Besides his larder service, he had to keep prisoners taken for offences in the neighbouring forest of Galtres,[1] to keep the king's corn measures, and to make distraints in the city of York for debts due to the king and sell the chattels which he had taken.[2] He was also said to be alderman of the minstrels. Though he was inadequately remunerated for his services by the land attached to the sergeanty, he was amply compensated by cash payments and rich perquisites. He received 5*d*. a day from the king's purse; and from 1164, at least down to the fourteenth century, this payment of 5*d*. a day or £7. 12*s*. 1*d*. a year is charged against the farm of the county.[3] He and his predecessors were also entitled to the following:

'On Saturday for every window of bakers selling bread a loaf or a halfpenny; of every alewife a gallon of beer or a halfpenny; for every butcher's window a pennyworth of meat or a penny; of every cartload of fish at the bridge over the Foss 4*d*. worth of fish, for which they paid 4*d*. at the price paid at the water's edge; similarly of every horse-load of fish 1*d*. worth of fish, for which they paid.'[4]

They received besides 4*d*. for every distraint they made. The position was, therefore, a lucrative one.

If but an insignificant landed estate pertained to the office of larderer at York, none at all appears to have been attached to some other offices held in sergeanty. This is evident from the history of the keepership of the king's houses at Westminster and of the Flete prison.[5] These offices were claimed in 1197 by Nathaniel of Leveland and his son Robert as their inheritance from the conquest of England. It is not improbable that this prescriptive title was justified so far as it relates to the keepership of the king's houses, though the available evidence can only prove that Nathaniel's father, Geoffrey the engineer (*ingeniator*), was in possession of it in the time of Henry I. From 1164 it seems to have been held together with the keepership

---

[1] An inquisition of 33 Edw. I (1305) mentions a house at York among the possessions of Philip the larderer called *prisona Lardinarie* (*Cal. Inq. p.m.* iv, no. 267).

[2] 'Per servicium custodiendi gayolam et vendendi averia, que capta sunt pro debitis domini regis' (*Book of Fees*, p. 356). Cf. *C.R.R.* i. 384.

[3] *P.R. 11 Hen. II*, p. 46; *14 Hen. III*, p. 267; *Cal. of Fine Rolls*, i. 525.

[4] *Cal. Inq. Misc.* i, no. 501. For his meat he was to have 'upper joints and loins' (*crura superiora et loynes*). *Cal. Inq. p.m. Henry III*, no. 753.

[5] The history of these offices has been traced in detail by C. T. Clay, *Eng. Hist. Rev.* lix (1944), 1 ff.

of the jail,[1] though the salaries attached to the two offices were distinct and separately accounted for on the farm of London and Middlesex. The former was paid at the rate of 7d. a day (£10. 12s. 11d. a year) and the latter at 5d. a day (£7. 12s. 1d. a year).[2] Nathaniel may have lacked the technical qualifications necessary for the performance of his duties, for during nearly the whole of the reign of Henry II the work was carried out by Ailnoth the engineer, who drew the full stipend of the office of keeper of the king's houses.[3] He may also have been the unnamed custodian of the Flete who after 1164 was in regular receipt of the salary attached to that post. However this may be, Ailnoth seems to have been responsible for supervising the whole area; he sees to the repairs of the king's chamber, makes preparations for the king's court, and is engaged from time to time in work on the jail, the bridge, and the quay at Westminster.[4] After more than thirty years' service, in 1189, Ailnoth was retired on a gratuity or pension of 50s.,[5] and the offices he relinquished were conferred by Richard I on Osbert, the brother of his much-favoured chancellor, William Longchamp. The claim of Nathaniel of Leveland was wholly ignored in this grant, and it was only after the fall of the family of Longchamp that he was successful in recovering his ancient inheritance, which he and his descendants continued to enjoy till 1558.[6] In 1204 the keeperships are said for the first time to be held by sergeanty.[7]

The Leveland family held of the archbishop of Canterbury a small property at Leveland in Kent—a half a knight's fee—for which they rendered knight's service.[8] Robert of Leveland was knighted by King John, went on crusade,[9] served as a

---

[1] The keepership of the Flete may have passed by marriage into the hands of the Leveland family (ibid., p. 5).

[2] These fixed stipends continued unchanged till recent times (Clay, op. cit., pp. 15, 17).

[3] The position of Ailnoth is difficult to explain. Mr. Clay suggests that Nathaniel appointed him as his deputy in view of his special qualifications.

[4] Ailnoth is twice mentioned in reference to work on the jail (*P.R. 19 Hen. II*, p. 91 and *31 Hen. II*, p. 44).

[5] Charged on the Wiltshire account of 1190 (*P.R. 2 Ric. I*, p. 118; Clay, op. cit., p. 7).

[6] Ibid., p. 15.   [7] Ibid., p. 8.

[8] *Red Book of the Exchequer*, pp. 471, 726.

[9] Robertus de Leveland, 'cui arma dedimus die Circumcisionis Domini (1202) et qui in crastino crucem assumpsit' (*Rot. de Liberate*, ed. Duffus Hardy, p. 25).

knight on juries for the grand assize,[1] and with other knights received advances of pay (*praestita*) when about to proceed on a campaign abroad.[2] But these chivalric services are quite independent of his sergeanty, for which he rendered the duties once performed by Ailnoth the engineer, namely, repairs to Westminster palace, preparations for the king's coming, custody of prisoners in the Flete, and the like.[3] For these he received the customary fixed stipends. The anomalous position of this sergeanty seems to have puzzled the jurors who made an inquisition on the property of Ralph de Grendon of Leveland in Edward I's time. They reported that

'he held the serjeanty of keeping the prison of Flete and the king's manor of Westminster and used to receive yearly from the king's money by the hands of the sheriffs of London £18. 5s. 8d. for keeping the said prison and manor, and for repairing the bridge of Flete, and also 100s. yearly from the free tenants of the soke of Flete by reason of the said serjeanty; but he held no lands in the county by reason thereof, and the jury know not whether he used or ought to receive anything else by reason of the serjeanty than the aforesaid moneys'.[4]

The office of keeper of Westminster palace and the Flete prison illustrates not only the strange heterogeneity of sergeanty, but also the almost casual way in which a sergeanty might come into being.

It was a feature of a number of sergeanties that the service was only demanded on the great feasts when the king, attended by his barons, wore his crown. At these ceremonial crown-wearings, held in the time of the Norman kings at Easter at Winchester, at Pentecost at Westminster, and at Christmas at Gloucester, serious matters of state were debated and important decisions were taken. But the king and his barons did not forgather merely for the discussion of state policy. Evidently the festive character of the season was not forgotten. Certain

---

[1] *C.R.R.* iii. 220; iv. 122.

[2] *Rot. de Praestito*, ed. Duffus Hardy, pp. 190, 205, 218; *Rot. Misae*, ed. Cole, p. 261. He subsequently joined the rebel barons and was among the knights captured at Rochester castle in 1215 (*Rot. Lit. Claus.* i. 241*b*).

[3] For his activities in his capacity of keeper of the houses at Westminster and of the Flete, see Clay, op. cit., p. 9 f.

[4] *Cal. of Inq. p.m.* ii, no. 356. The sum of £18. 5s. 8d. is clearly made up of the two stipends of keeper of the king's houses (£10. 12s. 11d.) and of the Flete (£7. 12s. 1d.) with an error of 8d. Property came to be attached to the sergeanty in the course of the fourteenth century. (Clay, p. 14.)

sergeants were assigned special roles to contribute to the merriment. It was on Christmas day, according to a late record, that the tenant of the manor of Kingston Russell in Dorset was required 'to count the king's family of chessmen in the king's chamber, and to put them back in the box when the king shall have finished his game';[1] it was on Christmas day, too, that Rolland, the tenant of Hemingstone in Suffolk, at one time performed his fantastic foolery.[2] Some of these aristocratic menials did not even perform an annual service. They allowed their work to be done by deputy, or to lie dormant only to be revived at the great ceremony of the king's coronation, which was preceded by an undignified contest among rival claimants who strove for the privilege of performing these offices and making off with the appropriate perquisites, the salt-cellars, the table-linen, or whatever it might be.[3]

The astonishing variety of the services, the extent to which they were actually performed, and the general characteristics of the tenure, may be aptly illustrated by the group of sergeants who are recorded as holding their lands in Oxfordshire. Among these there were a dispenser, a naperer, a man who prepared herbs, and a larderer; four were engaged with duties connected with the forest; there were several falconers; there was also a tailor, and there were three ushers.

The manor of Great Rollright was held at the time of the Domesday survey by a family who acted as dispensers; members of it were prominent at court and in the royal administration.[4] One of them, named Thurstan, was an itinerant justice in the reign of Henry II, and is revealed in a story told by the courtier and scandalmonger Walter Map as duly carrying out his functions as dispenser. He presented on bended knee to his fellow justice, Adam of Yarmouth, 'two royal cakes decently wrapped in a white napkin'.[5] The service of dispenser was often con-

---

[1] 'Ad narrandam familiam scaccarii domini regis in camera regis et ad ponendam in loculo cum dominus rex ludum suum perfecerit' (*Cal. Inq. p.m.* vii, no. 220 (1330)). An earlier member of the Russell family is said to have held by the service of being marshal of the butlery (*Book of Fees*, p. 92).

[2] 'Debuit facere die Natali Domini singulis annis coram domino rege unum saltum et siffletum et unum bumbulum' (make a leap, a whistle, and a fart): *Book of Fees*, pp. 136, 386, 1174, 1218.

[3] On the Coronation Services see Round, *The King's Sergeants*, ch. vi.

[4] See Round, *King's Sergeants*, pp. 186 ff.

[5] *De Nugis Curialium*, English version by M. R. James, p. 266.

fused with napery service; and it was Thurstan's grandson, also named Thurstan, who unsuccessfully contested with Henry of Hastings the right of rendering this service at the coronation of Queen Eleanor in 1236 and of purloining the table-linen after the feast.[1] Among the Oxfordshire sergeants there was in fact a naperer. A certain Robert, who came to be known as 'le Napier', inherited from his wife the fee of one knight in Pishill which belonged to the honour of Wallingford. King John released him from the, knight service and changed his tenure into sergeanty for which he was required to render at the Michaelmas Exchequer one table-cloth (*nappa*) worth 3s., or 3s. in lieu of the table-cloth every year.[2] This he did; but we find the payment recorded in rather unexpected places. It appears regularly in lists of scutages. Thus for the third scutage levied in this reign the entry is 'Robert le Napier (renders account of) one table-cloth for the fee of one knight'.[3] The new tenure never quite shakes itself free from the old.

Land at Ludwell in the parish of Glympton was held by the service of preparing herbs for the king at Woodstock,[4] or, as the service is defined in a final concord of 1195, the service which pertains to the *herbarium* of the king in the park of Woodstock.[5] It seems probable that this sergeant also belonged to the department of the dispenser, for in Henry III's reign we hear of a tenant at Ludwell (which is a very tiny hamlet, and unlikely to hold two sergeants) whose service was in the king's dispensary under Geoffrey the Dispenser.[6]

In the inquest of service taken in 1212 Adam de Mora is returned as holding 100s. worth of land of the gift of King Richard, and that he should be the king's *lardinarius*.[7] If he ever held this domestic office in the king's household, it was for a very short time. For when, in 1198, he received the land at Worton in Cassington to which the sergeanty became attached, he had already been for at least two years in the king's service as a falconer.[8] In 1199 he paid half a mark that 'it may be

[1] Round, op. cit., p. 224.
[2] *Book of Fees*, p. 117.
[3] *P.R. 4 Jo.*, p. 10.
[4] *Book of Fees*, p. 253.
[5] *Oxfordshire Fines*, ed. Salter, p. 53. Andrew Herborel, who held land at Ludwell in Henry III's reign, doubtless owes his surname to a connexion with this sergeanty (*Cal. Inq. p.m.* i, no. 256).
[6] *Book of Fees*, p. 830.
[7] *Book of Fees*, p. 104.
[8] *P.R. 10 Ric. I*, p. 194. Cf. *P.R. 8 Ric. I*, p. 290.

inscribed on the great roll . . . that the king wills that Adam shall hold his land of Worton with pertinances which King Richard gave him for his service by the service of falconry'.[1] This enrolment may be the official registration of the change of service. In the reign of John he was actively occupied in the business of falconry in the south-western counties.[2] In 1229, after Adam's death, the king granted 'the land that was Adam de Mora the falconer's', to Nicholas de Molis.[3] Though sometimes it is still entered in lists of sergeanties with the old service of falconry, both the service and the form of tenure were in fact changed. It became socage with the common render of a pair of gilt spurs.[4]

Besides Adam de Mora there were three other Oxfordshire tenants who held by falconry. Two of these, Robert Mauduit of Broughton Poggs and Robert of Liddington, whose lands lay round Bampton, were required to keep one falcon during the mewing or moulting season, and to deliver the bird to the king when it was able to fly.[5] This would seem at first sight an easy task to perform in return for a substantial property. But falcons needed careful treatment and an elaborate diet. King John sent instructions that they were to be given doves and pork, and chicken once a week;[6] and he wrote to the keeper of his favourite girefalcon, which he called Gibbon, that it must be provided with plump goats and good hens, and once a week the flesh of hares.[7] The service of keeping a falcon through the mewing season was not, therefore, just a sinecure. In the time of John the sport itself was conducted almost as a monopoly by one family, that of the Hauvilles.[8] No less than eleven members of this family were actively engaged in falconry. One of them, Walter, was given the manor of Bladon near Woodstock in return for this service, and during the first quarter of the thirteenth century he was very fully occupied with his business.[9]

[1] *P.R. 1 Jo.*, p. 229.
[2] Ibid., pp. 182, 188; *P.R. 5 Jo.*, p. 151.
[3] *Close Rolls, 1227–31*, p. 163.
[4] *Cal. of Charter Rolls*, i. 98; *Book of Fees*, pp. 1375, 1397.
[5] *Book of Fees*, pp. 103, 251, 589, 830, 1376, 1396.
[6] *Rot. Lit. Claus.* i. 118b.   [7] Ibid., p. 192.
[8] See Round, *King's Serjeants*, pp. 310 ff.
[9] *Book of Fees*, pp. 103, 253, 589. For his activities see *Rot. Misae*, ed. Cole, pp. 245, 251, 254; and ed. Hardy, p. 111.

A sergeanty attached to Stanton, which later came into the hands of the Harcourt family, is described as that of 'strewing fodder for the king's beasts at Woodstock, and of cutting and carrying a meadow of hay within the park at Woodstock'.[1] This park, we are told by William of Malmesbury,[2] was fenced in by Henry I for his menagerie, which is said to have included lions, leopards, lynxes, camels, and a porcupine sent to him by William of Montpellier. It is tempting to think that this strange sergeanty originated with the service of feeding Henry I's pet animals. However that may be, the king's beasts had long since died when about 1192 the sergeanty was given by Richard I to Henry de la Wade.[3] Henry de la Wade was in fact a falconer, and is shown by the records as engaged in this favourite sport of kings.[4] Henry, his son and heir, indeed held his land at Stanton by the service of keeping the king's falcons.[5] This younger Henry was also the king's cook, and for his good service in this capacity was rewarded in 1260 by the grant of another sergeanty at Bletchingdon.[6] The service consisted in providing the king with a dinner of roast pork (*veruta ad dinnerium regis, assam porci*) when he hunted in Wychwood forest.[7] It had passed through several hands before it reached those of

[1] 'Henricus de la Wade x libratas terre per servicium sternendi edere bestiis domini regis apud Wudestok et falcandi et levandi pratum infra parcum de Wudestok' (1212) (*Book of Fees*, p. 103).

[2] *Gesta Regum*, ed. Stubbs, ii. 485.

[3] In 1192 Henry de la Wade was paid 100s. of the king's gift 'until the king shall have assigned to him in some place a hundred shillings of rent' (*P.R. 4 Ric. I*, p. 294).

[4] *P.R. 32 Hen. II*, p. 178; *P.R. 4 Ric. I*, p. 294; *P.R. 5 Ric. I*, p. 133; *6 Ric. I*, p. 213.

[5] *Book of Fees*, pp. 253, 589. The history of the Stanton sergeanty is very confused. Henry de la Wade seems to have died in 1202, when his land was put into the custody of Geoffrey de Hauville, another falconer (*Liberate Roll*, ed. Hardy, p. 26). His heir Henry was still a minor in 1219 when William de Harcourt had the custody of his land. The old sergeanty of feeding the king's beasts appears to have passed before 1240 into the hands of the Harcourt family, for on an assize roll of the 25th year of Henry III it is stated that 'the jurors say that Richard de Harcourt holds Stanton in chief of the king by service of a third part of a fee of one knight and by sergeanty of collecting food in winter for the king's beasts and of mowing a meadow of the king under Evereswell and of carrying the hay . . . and Henry de la Waud holds eight librates of land in the same vill by the service of keeping the king's falcons' (*Book of Fees*, p. 1375; cf. *Cal. Inq. p.m.* i, no. 411).

[6] *Cal. Chart. Rolls*, ii. 29. The identification of Henry de la Wade the falconer and Henry de la Wade the king's cook is clear from an inquisition of Edward I's reign (*Cal. Inq. p.m.* ii, no. 620).

[7] *Book of Fees*, pp. 103, 253, 1374, 1397; *Cal. Inq. p.m.* i, nos. 213, 859.

Henry de la Wade the falconer. It had belonged in the twelfth century to Robert Fitz Nigel;[1] in the first half of the next century it was in the possession of the families of Grenville and Prestcote. The tenure, through changes and alienation, had become so confused that when in 1256 Henry III granted it to Master John of Gloucester his mason, he did so with the caution *quantum ad eum pertinet* and *salvo jure cujuslibet*.[2] Robert Fitz Nigel and Richard of Prestcote were local landowners[3] and may conceivably have performed the service. But the Grenvilles were substantial people in the West country,[4] and John of Gloucester was an ubiquitous hard-working man with a responsible and dignified office.[5] It is impossible to suppose that these men in person carried dinner to the king on the hunting-field. If the service were rendered at all, it was done by deputy.

On the roll of tenants-in-chief in Oxfordshire drawn up in 1212 there is an entry recording that Thomas son of Richard holds one carucate and he ought to be forester of Wychwood and to answer at the Exchequer for £7 annually.[6] This was Thomas of Langley (Langley, a hamlet in the forest where, it is said, there was a royal hunting-box) whose grandfather, Alan Rasur, had been forester-in-fee in the earliest years of the reign of Henry II.[7] Year by year the sum of £7 was accounted for at the Exchequer as the *census* of the forest of Wychwood (or,

[1] *Close Rolls, 1254–6*, p. 352. [2] Ibid.

[3] Robert Fitz Nigel made a grant to Godstow about 1190 (*English Register of Godstow Nunnery*, ed. Andrew Clark, p. 217). For Richard of Prestcote, see ibid., p. 218 and *Book of Fees*, pp. 447, 824, 840.

[4] *Rot. de Fin.*, p. 362. On the importance of this family in the west of England see Round, *Family Origins*, pp. 130 ff.

[5] The dignified position of the king's craftsmen in shown by an order to Peter de Rivaux, keeper of the wardrobe, to provide master Alexander the carpenter with a fur cloak 'such as the king's knights and master John of Gloucester, the king's mason, receive from the wardrobe' (*Close Rolls, 1256–9*, p. 177).

[6] *Book of Fees*, p. 103. The corresponding entry in the *Red Book of the Exchequer* (p. 455) wrongly calls him 'Robert son of Richard'.

[7] For the history of the family see Vernon J. Watney, *Cornbury and the Forest of Wychwood*, ch. 2. A relative, perhaps the father, of Alan Rasur accounted for the rent of the forest in Henry I's time (*P.R. 31 Hen. I*, p. 3). It was then £10, but after the detachment of the part of the forest round Stanton it was reduced to £7. The family also held land at Milton-under-Wychwood to which a small sergeanty held by Henry son of William was attached. This is described in 1219 as the service 'of finding a towell for drying the hands of the king when he hunts in the forest of Wychwood in the parts of Langley' (*Book of Fees*, p. 252). After 1219 it disappears from the records.

as it is designated in the Pipe Rolls, the forest of Cornbury). Thomas's duties were manifold. He carried out the king's instructions relating to the forest; distributed game and firewood to the king's friends; supplied timber for building; dealt with trespassers; and was generally responsible for forest management.[1] The custody of the other afforested area in Oxfordshire, which embraced Shotover and Stow wood, was in the thirteenth century in the hands of the family of Mimekan.[2] They held property at Headington in return for this service. Several letters on the Close Rolls, instructing them to provide timber and firewood to the recipients of the king's bounty, show that they attended in person to the administration of the forest.[3]

Oxfordshire provided two military sergeanties. One held by the family of Buffin at Nethercote near Tackley is of interest owing to its late creation. Round thought that the creation of sergeanties 'probably died out about the end of the twelfth century, or, at latest, in the reign of John'.[4] The sergeanty of William Buffin came into being in the reign of Henry III. Until 1236 or thereabouts he was holding his land by normal military tenure for the service of a quarter of a knight, and he paid scutage when scutage was demanded.[5] In 1240 the tenure is recorded as a sergeanty of the normal type of finding a sergeant in the king's army for forty days with doublet, steel cap, and lance.[6] Six years later it has already been partly alienated, and the service is divided among several tenants according to the size of their holdings.[7] Such divisions, like the subdivisions of knights' fees, must have added enormously to the difficulty of recruiting the army; and like fractional fees, they probably represent not physical service but money contributions.[8]

[1] For his activities see *Rot. Lit. Claus.*, passim.

[2] *Book of Fees*, pp. 252, 589, 1375, 1395. *Cal. Inq. p.m. Henry III*, no. 193.

[3] Peter Mimekan, for example, is ordered to supply ten cartloads of firewood to 'bishop John', a student at Oxford (*Close Rolls, 1227–31*, p. 265).

[4] *King's Sergeants*, p. 14.

[5] *Book of Fees*, p. 447. He was charged with scutage in 1203, 1205, 1206, and 1230.   [6] *Book of Fees*, pp. 830, 1173.

[7] Ibid., pp. 1173, 1215, 1397. The location of the sergeanty is wrongly stated in some of these entries.

[8] In one text two brothers Thomas and William Buffin are said to render the service alternately (*ad invicem*): *Book of Fees*, p. 1173. In another it is stated that each tenant shall render the service *pro parte sua secundum quantitatem tenementi sui*. The name of John 'ie Eskermisour' among the tenants may suggest that he performed the service for the whole group (ibid., p. 1173).

The service rendered by Robert Fitz Alan for his land at
Nether Orton near Cassington is obscure. When it is first mentioned it is described as the service of carrying the banner of
the people proceeding along the sea coast.[1] Next the banner
becomes the king's banner, and it must be borne within the
four ports of England;[2] but there is nothing to indicate which
ports were meant. By Henry III's time the service is defined
as that of carrying the banner of all the infantry of the hundred
of Wootton;[3] by 1242 it has become merely that of standard
bearer in the king's army for forty days at his own cost.[4] The
vagueness of the language used by the local jurors to describe
this sergeanty implies that the tenant of this little hamlet was
seldom called upon to perform the service.

The history of the king's tailor is perhaps better authenticated
than that of any of the Oxfordshire sergeants. Henry II granted
to Robert of St. Paul, his chamberlain, land at South Newington, which had belonged to Eschorsan *scissor* in the time of
Henry I. This Eschorsan may therefore be identified with the
tallator of the *Constitutio Domus Regis*, who was in the department
of the chamberlain and took his meals in the king's house.
Robert of St. Paul married Emma of Northampton, who
received this land at South Newington as her dower.[5] In the
inquest of 1212 she appears as holding it in sergeanty by the
service of tailoring the king's clothes; in 1219 the service is
defined as that of cutting the linen clothes of the king and
queen. After her death the land was escheated to the Crown,
and in 1227 it was granted to William the king's tailor. No
record has survived to inform us whether Emma herself actually
snipped and stitched. But there is abundant evidence of William's activities as a tailor in the later years of the reign of
King John and the early years of Henry III. Detailed tailor's
bills and orders for clothes are recorded on the Misae and the
Close Rolls.[6] He made robes for the king and queen and also

[1] 'servicium portandi baneram populi prosequentis per marinam' (*Book of Fees*,
p. 11).
[2] 'ferendi banarium domini regis pedes infra iiii portus Anglie' (ibid., p. 104).
[3] 'baneriam omnium peditum hundredi de Wotton' (ibid., p. 253).
[4] 'vexillum in host' regis per xl dies' (ibid., p. 830).
[5] Robert of St. Paul is described as *tailliator* in 1199 (*Rot. de Fin.*, p. 21). The
descent of this sergeanty has been traced by Round in *King's Sergeants*, pp. 257 ff.
and in *Rotuli de Dominabus* (Pipe Roll Society, vol. xxxv), pp. 22–3, and 23, note 1.
[6] For references see Round, *King's Sergeants*, pp. 261 ff.

for favoured members of the court circle; he repaired the crown and the coronation regalia for the king's coronation at Westminster in 1220. The land at South Newington he held by the render of a pair of scissors.[1] Renders of this kind were not necessarily paid directly into the Exchequer or the Wardrobe. They were often disposed of by the king as presents to his friends. In a letter preserved on the Memoranda Roll of 1230[2] the king informs the barons of the Exchequer that

'William our tailor by our command as of our gift has delivered to Ralph son of Nicholas one grey cloak, to Walter of Evermue one sore sparrowhawk, and to Henry Helion one pair of scissors, which he should have paid to our Exchequer at Michaelmas, and therefore we command you that you cause him to be quit.'

The grey cloak and the sore sparrowhawk were the respective rents for two other properties held by William the tailor, a house called Chapmanneshalle at Winchester[3] and a parcel of land in Kent.[4] In 1241 the land at South Newington passed by the gift of William the tailor in frankalmoign to the Hospital of St. John outside the East Gate of Oxford. But although it was granted in frankalmoign, the service of the old sergeanty was not forgotten; and the brethren of the hospital were themselves required to furnish the pair of scissors at Christmas, a further illustration of the practice, to which I have already drawn attention, of attaching a forinsec service to land held in frankalmoign.[5]

Of the Oxfordshire sergeants there remain three ushers, each of whom had special duties assigned to him. Hugh of St. Martin, who held land at Lillingstone Lovel, was usher of the king's hall, or, it is sometimes said, of the king's chamber.[6] Already by 1236 the service was restricted to the solemn occasions 'when the king wears his crown' or 'at the principal feasts'.[7] The usher service of the family of de la Mare, which

---

[1] 'per servicium reddendi quasdam forffices ad warderobam domini regis' (*Book of Fees*, p. 589).

[2] Pipe Roll Soc., N.S., vol. xi, p. 83.

[3] *Book of Fees*, p. 77; Round, op. cit., p. 258.

[4] *Cal. of Inq. p.m. Hen. III*, no. 183.

[5] *Cartulary of the Hospital of St. John the Baptist*, ed. Salter (Oxford Hist. Soc.), ii. 394.

[6] *Book of Fees*, p. 103. *hostium aule* (ibid., p. 253); *hostium camere* (ibid., p. 589).

[7] Ibid., pp. 589, 1397.

held property at Alvescot and Middle Aston in Oxfordshire besides lands in Gloucestershire and Wiltshire, is likewise spoken of variously as of the hall and of the chamber.[1] The special duty once assigned to them was that of looking after the harlots who follow the king's court.[2]

The usher of the Exchequer held his office by a sergeanty attached to the manor of Aston Rowant.[3] In the reign of Henry III it was stated that Roger de Scaccario and his ancestors had held their land here by the service of keeping the door of the king's Exchequer.[4] His ancestor, it may be assumed, was that Roger of Wallingford to whom Henry II about 1156 granted the *ministerium de magistratu hostierie de scaccario*.[5] Besides keeping the door this usher was responsible for conveying summonses issuing from the Exchequer throughout England. In an inquisition of 1271 (following the death of the second Roger) the office is defined as that of 'being grand usher of the king's Exchequer, usher of Jewry, and crier before the justices of the king's bench and the rest of the justices of eyre, for all pleas throughout the realm of England'.[6] Evidently he had an important part to play in the administrative work of the kingdom. But by the middle of the thirteenth century he was employing deputies to do it for him.

This brief survey of the sergeanty tenants who held their lands in Oxfordshire illustrates the remarkable diversity of the obligations attached to this tenure. Some, like William the tailor, Thomas of Langley the forester, or Walter de Hauville the falconer, were actively engaged on their appointed tasks; others, however, seem to have done little or nothing in return for their land. The fact that the form of service was frequently changed seems to indicate that the tenant was often ill suited

---

[1] *hostium aule* (ibid., p. 251); *ad hostium thalami regis*.(ibid., p. 830).

[2] 'per serganteriam custodiendi meretrices sequentes curiam domini regis.' Ibid. p. 253. For this and similar sergeanties see Round, *King's Sergeants*, pp. 96 ff., 108 ff.

[3] *Book of Fees*, pp. 1376, 1395.

[4] Madox, *Hist. of the Exchequer* (ed. 1711), p. 720, where the descent of the office is traced. Cf. also *Dialogus de Scaccario*, ed. Hughes, Crump, and Johnson, p. 23 f.

[5] *Cartae Antiquae* (Pipe Roll Soc., N.S., xvii), no. 277. Delisle, *Recueil des Actes de Henri II*, Introduction, p. 517, and vol. i, p. 190, gives the date 1156–8. But as Eyton (*Itinerary*, p. 19) notes, Roger was already in 1156 in receipt of a livery as 'Ostiarius de thesauro' (*P.R. 2 Hen. II*, p. 4).

[6] *Cal. of Inq. p.m.* i, no. 763. Cf. *Close Rolls 1264–68*, p. 496.

to perform the particular service assigned to him and was permitted to render another more to his taste and capacity. Sergeanties were hereditary, and would often pass into the hands of heiresses. The services might well be beyond the competence of an average woman. The manor of East Garston in Berkshire, for example, was part of the lordship of Kidwelly, held since before the middle of the twelfth century by Maurice of London and his successors.[1] It was a sergeanty, and the service required from the tenant was to lead the king's army with his standard and all his people through the midst of the land from Neath to Loughor when the king or his chief justice shall come to the neighbourhood of Kidwelly.[2] For fifty-five years, from 1219 to 1274, this lordship was in the possession of a lady, Hawise of London, who can hardly have been expected to perform this martial exercise during a stormy period of Welsh history. It may well have been to meet this sort of difficulty that early in Edward I's reign the service was changed to that of 'finding one armed horseman with his horse barbed with iron during the whole time the king shall remain with the army in the land of Kidwelly'.[3]

Sergeanty was already by the beginning of the thirteenth century somewhat antiquarian and out of date. It was rapidly outgrowing its usefulness. It had ceased to be an economical method of getting work done. Services were now being paid for or done by deputy; done inefficiently or not done at all. In the middle of the century we read 'the sergeanty of Henry of Monmouth in Leckhampton, for which he ought to be the king's cook, is changed into another service because the said Henry does not do the said service'. In fact it was commuted for a rent of a shilling a year and a fraction of knight service.[4] Henry of Monmouth held another sergeanty at Marden in Herefordshire to which other duties were attached. This, we are told, is also changed 'because the service is worn out' (est debile).[5]

Both the sergeanties of Henry of Monmouth were still intact

---

[1] V.C.H. Berkshire, iv. 247.

[2] Book of Fees, p. 863. Cf. Cal. of Inq. p.m. ii, no. 51, where the service is described as that of leading 'the vanguard of the king's army whenever he shall go into West Wales with his army, and the rearguard in returning'.

[3] Cal. of Inq. p.m. ii, no. 477 (1283). It was then again in the hands of an heiress who is stated to be an infant, aged one.

[4] Book of Fees, p. 1187.   [5] Ibid.

when they were thus changed into rents and knight service. It was an ancient rule of sergeanty tenures that they could not be divided among coheirs or be alienated.[1] The idea underlying these restrictions was that the service was an individual service which could not be shared among several tenants. Though attempts were made even in John's time to prevent division,[2] the rule was fast falling into abeyance. Early in the thirteenth century sergeanties were being partitioned among heiresses, and alienations were made on a large scale. It was to set matters in order that in the middle of Henry III's reign a number of commissions were set up to investigate the whole position. This inquiry resulted in what is known as the arrentation of Robert Passelew, the deputy treasurer.[3] The evidence reveals the chaotic state into which this class of tenures had fallen. So far from maintaining the principle that sergeanties cannot be alienated, we often find that they have been split into numerous small holdings. The small sergeanty of Richard of Pirie in Cirencester, for example, has been alienated into particles—*alienata est per particulas*; it has been broken into twenty-four little properties.[4] The general result of the inquiries made between 1246 and 1250 was that a number of sergeanties came to an end. They were commuted into money rents, sometimes together with a fraction of knight service.[5] In some cases the old service was retained, and the responsibility for its performance was thrown on the original tenant or on one of the alienees. As a reality, sergeanty was rapidly falling out of the scheme of feudal society.

[1] In 1205 it was stated in the king's court that 'non potuit vel debuit aliquis sergantiam delacerare vel aliquo modo alienare'. *C.R.R.* iv. 35.

[2] See, for example, *P.R. 7 Jo.*, pp. 238, 274.

[3] See *Book of Fees*, pp. 1163 ff., and Kimball, *Sergeanty Tenure*, pp. 229 ff.

[4] *Book of Fees*, pp. 1188, 1248 f.

[5] The law appears to have adopted a different attitude towards complete and partial alienation. In the former case the land was taken into the hand of the Crown. So on the Yorkshire Eyre of 1218–19 it was adjudged that because certain land was the king's sergeanty 'and William Mala opera could not sell the whole (*totam*) sergeanty nor could another buy it without the king', it should be taken into the king's hand (Selden Soc., vol. lvi, no. 1147). Cf. *Book of Fees*, p. 1168: 'Peter of Malling gave that sergeanty wholly (*integre*) to Robert de Bosco, so it was determined that it should be in the king's hand.' A sergeanty, on the other hand, which was alienated *in parte*, was usually charged with rent and a portion of knight service.

V

AMERCEMENTS

ONE of the chief burdens which fell upon all men during this period was the payment of amercements. A man for some reason or another, often quite innocently, gets himself into trouble with the authorities. He is adjudged *in misericordiam*; he is in the king's mercy; he is amerced. Maitland[1] thought that most men must have expected to be amerced at least once a year, and he is hardly exaggerating. The very inevitability of amercement is sufficient justification for regarding it among social obligations. These amercements fell largely on the village communities, on the class of small freemen and villeins. The sums will often appear to be very trifling, though in the king's court they never dropped below half a mark, that is, 6s. 8d. But to appreciate the full weight of this burden we must arrive at some idea of the relative value of money then and now, and of the ordinary man's ability to pay.

Let us consider first the relative value of money. It has often been suggested that we can arrive at a fairly accurate estimate of money values by multiplying medieval figures by a given number. Sir Thomas Duffus Hardy,[2] writing in 1833, proposed multiplying by fifteen. Dr. Round,[3] writing in 1912 in one of his valuable introductions to the Pipe Rolls, tentatively suggested multiplying by forty. Dr. Coulton,[4] writing independently in 1934, and with reference to prices in the later Middle Ages, adopted the same index figure as Round. But not only did prices differ considerably in the twelfth and fourteenth centuries, the periods with which Round and Coulton are concerned, but there is a still greater discrepancy between prices in 1912 and prices in 1934. An arbitrary system of this kind is apt to be very misleading. We shall be on surer ground if we bear in mind the average rate of wages earned by the lower classes in the twelfth century, and compare it with the rate in force at any particular time. I would not take for the purpose

[1] Pollock and Maitland, *Hist. of Eng. Law*, ii. 513, 519.
[2] *Rot. Lit. Claus.* (Record Commission), i, Introd., p. xlv.
[3] Pipe Roll Society, vol. xxxiii, Introd., p. xxx.
[4] *The Meaning of Medieval Moneys*, Historical Association Leaflet, No. 95.

of this comparison the inflated prices obtaining at the present time, for this would give a very unreal idea. The whole annual revenue of the Crown in the twelfth century would possibly keep the British war effort going for about twenty seconds. Let us take the year 1938, when the average agricultural wage was approximately 35s. a week.[1] Now the normal wage of the working-classes in the twelfth century may be taken at one penny a day. To take a few examples from the accounts of the later years of Henry II's reign, we find a foot-soldier, a porter, a carter, a chaplain, the king's fool, all earning 1d. a day, or 30s. 5d. a year. There were of course exceptions to the general rule. The under-servants in the king's household in the time of Henry I were given 1½d. a day. At the close of the century, perhaps owing to difficulties of recruitment and the expense of equipment, soldiers and sailors were getting 2d. Huntsmen were a particularly favoured class, and could expect to get 2d. a day with an allowance of ½d. for each hound under their charge. Nevertheless, 1d. a day was the normal, and also the minimum wage. William, King John's bathman, was, it is true, in receipt of only ½d. a day, but he had a bonus of 5d. every fortnight when the king took his bath, and probably other 'extras'.[2] If these wages are compared with an average agricultural wage of 35s. a week, that is, sixty times as great as the usual twelfth-century wage, the sums of money which we shall quote will not seem so trifling. Even the smallest amercement taken in the king's court, half a mark, would on this reckoning correspond to a fine of £20 in modern times, a fairly crushing fine to the ordinary labourer.

The average man of the working-classes was then earning 1d. a day or 30s. 5d. a year. What were his capital assets? This is not so easy to discover, and it will be necessary to descend to the criminal classes to find any sort of answer. On two occasions, in 1166 and again in 1176, Henry II made a thorough round-up of the criminal population. On the first of these occasions the justices, acting on the instructions issued in

---

[1] The actual figure estimated by the Ministry of Agriculture was 34s. 7d. a week (*Ministry of Labour Gazette*, April, 1939).

[2] *Rot. Misae* (ed. Hardy, pp. 115, 137). In Henry I's time the *Aquarius* had double rations, 1d. when the king was on his travels for drying his clothes, and 3d. when the king took his bath ('Constitutio Domus Regis' in *Red Book of the Exchequer*, p. 811).

the Assize of Clarendon, condemned 574 named criminals; in the second, following the Assize of Northampton, 705 men were condemned.[1] The personal property of these, whether· they fled from justice and were outlawed, or stood their trial and were condemned under the ordeal, was confiscated and sold for the benefit of the Crown. The sums so received appear on the rolls under the heading 'the sheriff renders account of the chattels of fugitives and of those who perished (that is, failed) in the ordeal of water', or 'of the chattels of fugitives and of the defooted (*expedatorum*)' because the penalty prescribed by these assizes for those who failed in the ordeal was the loss of a foot. Now the total value of the chattels of the first batch of criminous peasants amounts to £312, which gives us 10s. 11d. as the average value of the personal property of each individual. The property of the second batch amounts to £343, making an average of 9s. 8d. Though the individual sums vary considerably, from 6d. to 70s., I think we shall not be far wrong if we assume that the goods and chattels, the net personalty, of the ordinary peasant was about 10s.

Before discussing the burdens imposed by legal procedure on the lower classes of society I wish to draw attention to another class distinction. Glanvill marks a distinction between the mode of trial for criminal offences which the free and the unfree had to undergo. The free carry the hot iron, the unfree go to the ordeal of water.[2] Now it seems that the aristocracy were seldom required to submit themselves to either of these ordeals. We read, for example, on the Pipe Roll of 1168 'for the cost of one knight hanged and for the ordeal of one man who failed and was hanged, 8s.'[3] It appears from this statement that there was no necessity for divine proof in the case of persons of quality; they could be hanged without ceremony. The system of ordeals was concerned with the proletariat. Of the hot iron we hear little. Both the Assizes of Clarendon and Northampton prescribe the water ordeal for all cases; and the evidence of the Pipe Rolls shows that this was followed in all counties except Hampshire, where in 1176 *judicio ignis* is substituted for *judicio*

[1] In some counties the details are insufficient for statistical purposes and have to be omitted. They give only the total sum received from the sale of chattels without the number of criminals condemned.

[2] xiv, cap. 1.

[3] *14 Hen. II*, p. 188.

aquae in the lists of those who failed to pass the test.[1] In fact
it seems to have been rarely used except for women offenders
who invariably carried the hot iron. This is very clearly
brought out in a Cornish case of 1201, where seven persons,
six men and one woman named Matilda, were implicated in
a burglary. The verdict was 'Let the males purge themselves
by water under the Assize, and Matilda by ordeal of iron'.[2]
Whether Ranulf Glanvill or Hubert Walter was the author of
the famous law-book is, in this question, unimportant. Both
were experienced justices, and Glanvill, carrying with him the
instructions issued at Northampton, had travelled on the
northern circuit in 1176, when many hundreds of persons were
sent to the ordeal of water. Are we to assume that the author
of the treatise was merely speaking theoretically, or that all
these men were in fact villeins? Or, to put it the other way,
that no free men committed crimes? It seems probable that
this distinction between the free and the unfree was not strictly
applied. A man would not stand on his privilege as a freeman
in order to get himself tried by iron, which was a far severer
test than the ordeal of cold water.

In a sense it may be said that the arbitrary amercement
replaced the old English system of *bot* and *wite*, a system by
which crimes were attoned for by money payments to the
wronged person and to the State according to a carefully and
minutely graded tariff to meet offences of various kinds. The
principle of amercement was, however, essentially different.
The idea of compensation to the injured party was altogether
absent. Damages were only awarded for wrongful disposses-
sion, for disseisin, or for infringement of rights or nuisances
(*nocumenta*), such as obstruction of water courses, interference
with landmarks, or the like.[3] Nor yet were amercements
generally imposed as punishments for crimes committed. Only
minor misdemeanours (*transgressiones*), such as selling wine or
cloth contrary to the regulations, were punished by a money
fine. Sometimes an unusual event gave rise to a crop of
amercements. Over £25 was collected by the justices in 1200

---

[1] *P.R. 22 Hen. II*, p. 197.
[2] *Select Pleas of the Crown*, ed. Maitland (Selden Soc.), no. 12.
[3] The assize of novel disseisin was usually the appropriate action in such cases.
Cf. *C.R.R.* iii, pp. 90, 132, 139, 214–15; and below p. 90.

from a group of Lincolnshire men 'for putting a woman unjustly
on the tumbrel',[1] and more than double that sum was taken the
next year in the same county *pro crasso pisce*, for a fat fish,.
presumably a whale or porpoise, which were a royal prero-
gative.[2] No man must take liberties with the royal fish; he
would be amerced. But these were relatively trivial misdeeds.
Crime, or at least serious crime, was punished not by amercement but by death, mutilation, or outlawry. The vast majority
of amercements imposed by the justices in eyre were not for
sins of commission but for sins of omission; for failure or neglect
of the public duties compulsorily thrown upon the community
or the individual, or for mistakes, which were often inevitable,
in the course of procedure.

Common law was growing more rapidly than the machinery
of justice and police. No doubt the anxiety of the justices to
collect as much money as possible for the Exchequer was in
part responsible for the great number of amercements imposed
on a judicial eyre, but the primary cause must be sought in
defects of legal procedure and in the lack of a professional police
force. The burden of detection of crime and of the capture of
criminals, and of bringing them to justice, rested almost entirely
on the unassisted and unremunerated labours of the local community of peasants and townsmen, vested with little or no
authority or with the means to carry out the obligations imposed upon them.

An examination of an eyre roll leaves one with the impression
that it was almost impossible, however much he might strive
to do his duty, for a man to escape amercement at some stage.
It is easy to understand the dread which the coming of the
justices inspired, and why a town like Leicester in 1180 would
pay a substantial sum of 80 marks 'that they may be quit of
murder fines and common pleas on this visit of the justices',[3]
or that in 1202 the city of Lincoln, when there were 11 charges
of murder, 13 of robbery with violence, 4 of assault and wounding, 5 of rape, 2 of forgery, and 4 other breaches of the peace,
paid 100 marks that they might not be interfered with ·in their

[1] *P.R. 2 Jo.*, p. 77 f.
[2] *P.R. 5 Jo.*, pp. 121 f., 98. According to Bracton, some thought that in the
case of the whale the royal right should be satisfied if the king had the head and
the queen the tail (f. 120*b*).
[3] *P.R. 26 Hen. II*, p. 101.

pleadings, and with another sum of £100 compounded for licences to settle matters out of court and for their misdemeanours.[1] Chartered boroughs, like Leicester and Lincoln, had the necessary power and organization to drive a bargain, and could thus rid themselves of much of the burden and annoyance which an eyre occasioned. But the hundreds and townships had no such organization: they had to go through with it; and the serious results are reflected in the long lists of amercements which were drawn up at the close of a session of the justices.

To illustrate the extent of these burdens we will take the Lincoln assize of 1202, the rolls of which were finely edited by Mrs. Stenton for the Lincoln Record Society in 1926. The justices had visited Lincolnshire in the two preceding years, in 1200 and 1201. The crimes, therefore, with which the justices, Simon Pattishall and his associates, had to deal, were probably of recent occurrence. And yet the volume is enormous. There were approximately—it is difficult to classify and disentangle some of the cases—114 cases of homicide, 89 of robbery, generally with violence, often with great violence, 65 of wounding, 49 of rape, besides a number of less serious breaches of the peace.[2] Now what punishments were inflicted for all this mass of brutal and squalid crime? Two men were hanged; 16 men, who had fled and escaped capture, were outlawed; 11 others had taken sanctuary in a church, and all that the law could impose on them was a sentence of outlawry; 9 criminal clergymen were handed over to the court of Holy Church, where, no doubt, they were dealt with lightly. Eight cases were sent for trial by ordeal, 3 by fire and 5 by water. The ordeal of hot iron was, as I have said, a severe test, and it was often applied with gross unfairness; for in two of the three cases at the Lincoln assize the option of whether the accuser or the accused should carry the iron was given to the accused, who naturally enough opted for his accuser. The result was that neither did so, for they both put themselves in mercy and were duly amerced.[3] The ordeal of cold water was a far easier test to surmount. Only one case of failure is recorded; and the

---

[1] *Linc. Ass. Rolls*, nos. 1017, 1095; *P.R. 4 Jo.*, p. 240.

[2] These figures are based on an analysis made by I. L. Langbein in the *Columbia Law Review*, xxxiii (1933), 1337, note 24.   [3] *Linc. Ass. Rolls*, nos. 595, 843.

man thus condemned by divine judgement was presumably outlawed, for his chattels were confiscated.[1] The Assizes of Clarendon and Northampton had prescribed that a man of bad repute, although he succeeded in the ordeal, must abjure the kingdom. Yet even this salutary rule was not enforced on the Lincoln assize of 1202.

'Letitia of Clixby' (we read on the plea roll[2]) 'appealed Hugh Shakespeare that he came to her mother's house and bound her and her mother, and robbed them in the peace of the lord king and basely. And the jurors suspected him of the crime, and the four nearest vills likewise suspected him, and they declare that he was of evil repute. And so let him purge himself with water. And he waged his law and was saved. And he gives to the lord king two marks that he need not abjure the kingdom but find good pledges for his faith.'

Evidently the amassing of money for the treasury was considered of greater importance than ridding the country of a scoundrel.

It was the general rule that an accuser should follow up his appeal by offering to prove it by his body, that is, by a duel. But there is more talk than fighting. Often one of the parties would submit that he was maimed and could not fight, or that he was past the age of fighting—over sixty. Both these excuses were allowed by the law. Only in two cases on this eyre did the court order a combat. In one of these the issue is uncertain. Here is the record of the other:[3]

'Ralph the blacksmith appealed Ralph son of Jordan that he in the peace of the lord king and basely assaulted Agnes his wife and robbed her of five shillings and three rings and two silver brooches, and he gave her a wound on the head, and besides that he broke the window of her house, and this he offers to prove against him in the consideration of the court as of his sight and hearing. And Ralph defends the whole. And it is considered that there be a duel between them, because the jurors bear witness that he raised the hue and cry and that he exhibited a little wound. They came armed on Tuesday after the octave of the feast of the apostles Peter and Paul at Lincoln. Afterwards they came and put themselves in mercy.'

This is a common ending to a trial by battle. They did not fight at all. The duel was a violent and often a mortal form

---

[1] Ibid., no. 588d. [2] Ibid., no. 855. [3] Ibid., no. 638.

of proof. A late description of it[1] informs us that if the com-
batants broke their weapons they must still go on fighting 'with
their hands, fists, nails, teeth, feet, and legs'. It is not surprising
that men withdrew their accusations or sought leave of the
court to compromise their quarrels.

Two capital sentences and some twenty or thirty sentences
of outlawry are then the only physical punishments inflicted for
about 430 cases of more or less serious crime. On the whole
a disappointing result. For the rest it is a matter of pounds,
shillings, and pence. This money, however, is not collected
from condemned criminals but from men who have either in
one way or another failed in their public duties or have made
mistakes in their pleading or in the course of procedure. The
roll of amercements on this Lincolnshire assize (excluding the
city of Lincoln which, as we have seen, compounded for its
amercements) totals up to £633. 15s. If we deduct the amount
taken for selling wine contrary to the assize, a few fines for
special privileges, and the value of the chattels of outlaws, we
arrive at the figure £527. 12s.

If we analyse this amercement roll more closely we notice
first that a considerable majority of the Lincolnshire wapentakes—21 out of 32—are charged with murder fines. There is
still much that is obscure about this murder fine, the *murdrum*.
But we may probably accept the view, propounded by the
author of the *Dialogue of the Exchequer*,[2] that it originated with
the Conqueror, who desired to protect his Norman countrymen
from the violence of the native English. These, we are told,
would often cut the throats of the foreigners in woods and
lonely places. Consequently the fine was not imposed if the kin
could prove to the satisfaction of the justices that the slain man
was not a Norman but an Englishman. The lawyers further
insist that the fine can only be taken in cases of secret homicide,
nullo praesente, nullo sciente, nullo audiente, nullo vidente, as Bracton
says.[3] The justices, however, do not seem to pay the slightest
attention to these rules. They would impose the fine when they
knew perfectly well who had committed the crime. Ten of the

---

[1] *Gregory's Chronicle*, ed. Gairdner (Camden Society), p. 200; cited by H. W. C.
Davis in *Eng. Hist. Rev.* xvi (1901), 730. Bracton (f. 145) considered the front
teeth as a valuable asset in the combat: 'hujusmodi dentes (praecisores) multum
adjuvant ad devincendum.'

[2] i. 10.   [3] f. 134b.

cases on the Lincoln roll are of such a kind; the murderer is known and has fled from justice. It is also worthy of remark that only in a single instance was Englishry presented.[1] These murders arose from quarrels among ordinary villagers; and yet only one family came forward to claim the dead man as an Englishman, and so save the community of the hundred from the fine. These peasants were able, as we have seen, to produce elaborate pedigrees when the question of their personal status was involved. Taking full account of the mixture of races through intermarriage upon which Richard Fitz Nigel writing some twenty years before dilates,[2] it is difficult to believe that all these murdered men were Norman or even partly Norman. It is more probable that presentment of Englishry was not encouraged by the itinerant justices on this visit. The custom varied in one district and another; and no doubt too the practice of judges differed. An examination of the rolls of the west-midland eyre of 1221 reveals a large number of presentments of Englishry.[3]

The amount of these fines was fixed very arbitrarily. The crushing sum of 46 marks or rather over £30, mentioned in the early law-books, seems never to have been exacted; but often they were sufficiently high to be a serious imposition. They varied enormously. On the Lincoln roll they vary from £10 to 13s. 4d. It is difficult to account for this diversity. They might of course be graded according to the brutality, the sordidness of the crime. At first we seem to get an inkling of this from the habit of the clerk of expressing his feelings in the margin of his record. He would repeat the word *murdrum* two, three, four, or even five times. When Roger son of Ailric murdered his wife, the clerk wrote *murdrum* three times in the margin, and the fine was quite properly assessed at the highest figure of £10.[4] But when he records that one Tubon was found dead in the wood of Evedon and no one knew who killed him, he writes the word four times, and for this crime and another

[1] No. 819.
[2] *Dialogus de Scaccario*, i. 10.
[3] In Worcestershire, for example, in 39 out of 66 cases of homicide it is stated that Englishry was presented; in 27 that it was not. Likewise there were a large number of presentations of Englishry in Gloucestershire. The Worcester rolls have been edited by Mrs. Stenton for the Selden Society, vol. liii, pp. 534–619. *The Pleas of the Crown for the County of Gloucester* were edited by Maitland in 1884.
[4] Nos. 942, 1066.

murder committed in the same district the wapentake is only called upon to pay the relatively trifling sum of £2. 13s. 4d.[1] We can only assume that the judges were in possession of more facts than have been preserved on the roll, and that they assessed the fine accordingly.

The king was accustomed to grant to churches and to favoured subjects charters relieving them of the burdens which normally fell upon land and its tenants. Thus, for example, on 25 January 1190 Richard I addressed to all that it might concern a writ in the following terms:

'Know that for our salvation and for the soul of our father King Henry and for the souls of all our ancestors we have conceded and by our present charter have confirmed to God and to the church of the blessed Mary at Lincoln and to Hugh bishop of Lincoln and to all his successors in free and pure and perpetual alms that all their lands and their men may be quit of the murder fine and of the pence pertaining to murder.'[2]

The possessions of the see of Lincoln stretched far afield, and we see the repercussions of this charter here in Oxfordshire. The hundred of Dorchester-on-Thames, which belonged to the see, had incurred a murder fine in 1188.[3] In 1190 there is still a small balance to be paid; but against the entry on the Pipe Roll is written 'pardoned by the liberty of the king's charter to the bishop of Lincoln'.[4] Not only churches, but barons and even some privileged towns, enjoyed similar immunities.[5] The result is indicated on the roll of amercements where, after the statement of the amount of the murder fine, the words *exceptis libertatibus* are significantly added. In effect, therefore, the incidence of the *murdrum* fell not on the whole hundred or wapentake, but on those parts which lay outside these great seignorial franchises.[6] It threw the burden chiefly on the smaller men.

---

[1] Nos. 761, 1048.   [2] *Registrum Antiquissimum* (Lincoln Record Society), i. 124.
[3] *P.R. 34 Hen. II*, p. 154.   [4] *P.R. 2 Ric. I*, p. 12.
[5] See Ballard, *British Borough Charters, 1042–1216*, pp. 150–1. Besides the towns there mentioned, Chesterton also had exemption. In 1199 Geoffrey Fitz Peter wrote to the sheriff of Cambridge and Huntingdon: 'We order you not to exact or allow to be exacted from the men of Chesterton the *murdrum* which they ought not to give and have not been accustomed to give' (*Memoranda Roll 1 Jo.* (Pipe Roll Soc., N.S. xxi), p. 36).
[6] A very different custom is suggested by the vague language of the early law-books (*Leg. Hen.* 91. 2; cf. *Willelmi Articuli*, 3). This is to the effect that if the

Nevertheless, the hundred, even omitting the franchises, might contain a fair number of persons to share in the fine. The burdens and responsibilities imposed on what is known as the frankpledge were far more serious. Owing to the lack of professional policemen, these groups of ten or a dozen men, composed mainly of peasants, were charged with the duty of looking after each other's behaviour. If one of their number has committed a crime, the rest must raise the hue and cry, follow him, and, if they can, take him. If they fail to catch him, they are amerced. If they succeed in bringing him before a court, their liability is generally at an end; and it is usually some friends or neighbours who now go bail for the criminal and guarantee that he will appear before the justices when they next come into the county. They may not come for some time; the case for one reason or another may be postponed. In the meantime the suspect may have escaped, and his pledges are at mercy. £95 was taken off these wretched Lincolnshire men at the eyre of 1202 for failure of pledge.

The position of an injured party was far from easy. Even after public indictment by a jury had become a regular part of legal procedure,[1] personal appeal continued to be the most common means of bringing a criminal to justice. On this Lincolnshire assize approximately 353 prosecutions were brought by personal appeal and 77 by jury presentment.[2] If the wronged person did not prosecute, he might be in trouble and be amerced; if he started on his prosecution and afterwards gave it up, he was likewise amerced; if he carried through with it, and the evidence was deemed to be insufficient, he was at

body is found on a manor, the incidence of the fine fell in the first instance on the manor itself. Though there is not evidence to show that this was at all a general practice, Robert of Courtenay successfully claimed it for his manor of Sutton (Courtenay) in Berkshire. 'He said in the presence of the barons of the Exchequer that if a *murdrum* fell on his manor of Sutton he ought to pay the amercement so that others who are outside that manor ought not to participate with him, nor conversely' (*Memoranda Roll 14 Hen. III* (Pipe Roll Soc., N.S. xi), p. 35). He made the claim, however, to secure freedom from participation in murder fines outside his manor. This is clear from a mandate recorded on the same roll to the sheriff ordering him to release Robert of Courtenay from a demand for the fine made on his manor of Sutton 'because the murder happened in the hundred of Sutton outside that manor' (ibid.).

[1] For a valuable discussion on the origin and development of the Jury of Presentment see Miss N. D. Hurnard, *Eng. Hist. Rev.* lvi (1941), 374.

[2] These figures are taken from the analysis of Langbein, loc. cit.

mercy for a false claim. Although in the twelfth century a man might employ an attorney in civil cases to do his work for him, this was not permitted in criminal cases. Here a plaintiff gets no professional help. If he makes some mistake in his pleading, if there is some technical flaw in his appeal, he is amerced *pro stultiloquio*, for silly pleading. Perhaps the most common ending of a suit was an agreement of the parties to settle the matter out of court. They would pay for leave to do this (*pro licentia concordandi*), and a jury would assess the amount that each should contribute.

The accusing jury can have been little more at ease than the plaintiff or the defendant. They were liable for amercement for concealing a plea or for not presenting a plea which they ought to have known about. They are also amerced for making a false or a foolish presentment. It was hard to steer a safe course between Scylla and Charybdis. The jurors might speak out, and be at peace with the king's judges; but in doing so they might arouse the enmity of powerful local men who could do them much harm at home. Such a situation probably lies behind an entry on the roll of 1180: 'The village of Lambourn renders account of two marks except eight jurors who spoke the truth.'[1] It is not surprising that men would pay a mark or so to escape serving on a jury.[2]

We are accustomed to praise the legal reform of the twelfth century and to regard it as the great legacy which Henry II contributed to our legal system. Ultimately it led to our common-law procedure of which we may be justly proud. It is, however, doubtful whether it worked well in its initial stages. It certainly favoured the criminal. Perhaps the rough justice meted out in Henry I's day was better suited to the times. There was little formality about the proceedings when, according to the well-known passage in the *Anglo-Saxon Chronicle*,[3] Ralph Basset and the king's thegns at Huncot in Leicestershire between St. Andrew's mass and Christmas 'hanged so many thieves as never were before, that was in that little while four-and-forty men'. The new procedure was little more than an expensive game of forfeits when the minimum stake was half a mark, 6*s*. 8*d*. It was also a compulsory game, a game at

---

[1] *P.R. 26 Hen. II*, p. 40.
[2] See, for example, *Lincoln Assize Rolls*, no. 632.
[3] *Anno* 1124.

which heads I lose, tails you win. In all these cases heard before
the justices at Lincoln there was no question of compensation
or damages to the plaintiff except for disseisins. You were
almost bound to come out of court poorer than you went in,
whether you were there as plaintiff or defendant, pledge or
juryman.

I have dwelt at some length on the question of the amercement of peasants in order to show that by the twentieth chapter
of the Great Charter this class obtained a very substantial
benefit. It runs, you will remember, as follows:

'A freeman shall not be amerced for a small offence except in
accordance with the measure of the offence; and for a serious
offence he shall be amerced according to the greatness of the offence,
saving his *contenementum*; and a merchant in the same way, saving
his merchandise; and a villein shall be amerced in the same way,
saving his *wainagium*, if they shall have fallen into our mercy; and
none of the aforesaid amercements shall be imposed except on the
oath of honest men of the neighbourhood.'

The clause is noteworthy as the only one which *eo nomine* mentions the villeins. It does in fact cover all classes of society
except the barons and the clergy who are dealt with in separate
chapters of the charter. Those scholars who have sought to
detract from the importance of the charter have interpreted
these technical words *contenementum* and *wainagium* in a sense
which would make the whole clause chiefly one of baronial
interest. So McKechnie in the first edition of his great monograph on *Magna Carta*[1] defined *contenementum* as 'tenement'
which it was obviously in the lord's interest to exclude from
amercement; *wainagium* was the villein's 'plough and its
accoutrements' without which the villein could not perform the
labour services due to his lord. Professor Tait with his accustomed thoroughness proved the incorrectness of these inter-
pretations,[2] and in his second edition McKechnie, with obvious
reluctance, adopted Tait's conclusions.[3] *Contenementum* means
a man's status, his social position. The freeman must not be
so heavily amerced that he cannot maintain himself in his
position in society. *Wainagium* of the villein is his tillage or

---

[1] Ed. 1905, pp. 343, 345.
[2] *Eng. Hist. Rev.* xxvii (1912), 720 ff.
[3] *Magna Carta*, 2nd ed. (1914), pp. 291 n. 1, 293 n. 6.

crops. The evidence on which Professor Tait relied belongs to
the period subsequent to the granting of Magna Carta. It can,
however, be corroborated by evidence from the period just preceding the charter. The word *wainagium* occurs four times in
the Curia Regis Rolls of the reign of John. In each case the
plaintiff complains of obstruction of his wainage, *de impedimento
wainagii*, with consequent damage which is assessed. Thus, for
example, Brian of Fallege is summoned to show why he did
not permit the prior of Newark (Surrey) to cultivate his land
in peace; and the prior goes on to complain that Brian took
away from that land corn to the value of 40*s*. and by the
obstruction of his tillage or cultivations (*de impedimento wainagii*)
he suffered loss to the value of 10 marks.[1] In these legal records
there can be no doubt that by wainage was understood tillage
or cultivations from which the villein earned his livelihood.
Occasionally the word is also used in the second sense which
Professor Tait attributes to it—that of crops. In an inquest
made by the order of King John concerning the services and
customs of the manor of Cirencester it is stated that all those
who brewed from their own wainage shall be quit of toll.[2]
Peasants could brew good beer from their barley crop, but they
could not conceivably brew with a plough. But whichever
meaning we adopt, whether tillage or crop, it amounts to the
same thing; it is the villein's means of livelihood that is protected against amercement. And this is, in effect, what is
reserved to each of the three classes mentioned in this chapter
of Magna Carta, the freemen, the merchants, and the villeins.
A little later, in fact, the same word *contenementum* is used for
the villeins as well as the freemen. In the Hundred Rolls we
are told that the serfs of Swyncombe in Oxfordshire must only
be tallaged 'saving their *contenementum*' and contribute to all
common aids according to their means.[3]

Chapter 20 of Magna Carta granted a substantial measure
of protection to the villein. But how, we may ask, could it be
enforced? Amercements by the same clause were to be assessed
by honest men of the neighbourhood. It may be that these
reservations were merely intended as instructions to the assess-

---

[1] *C.R.R.* iv. 44. Cf. also v. 209, 274; vii. 187, 206.
[2] *Trans. of the Bristol and Gloucestershire Arch. Soc.* ii. 297.
[3] ii. 758; Vinogradoff, *Villainage*, p. 177.

ing jury. But there is a passage in Bracton[1] to the effect that should the lord take away the villein's wainage, the villein has an action. The text here is not free from difficulties, and such an action cannot be proved from legal records. But it may well be that Bracton believed that at least in theory the villein had an action at law against his lord who brought him to economic ruin.

The word *misericordia* had a sinister meaning in the Middle Ages. It might spell ruin before it was mitigated by the twentieth chapter of the Great Charter. To be at the king's mercy was very terrible. Rich and poor were ruthlessly amerced and fined. Yet the ruthlessness of the Middle Ages may be exaggerated. Even King John is said on a plea roll to have been 'moved by pity'.[2] The honest men of the district who assessed the amercements evidently took account of the circumstances of the men at their mercy. Fines were sometimes altogether remitted on account of poverty even before 1215.[3] The ensuing civil war brought with it much misery and destitution. The rolls of the eyres which followed the return to normal conditions show that the judges were anxious to temper the wind to the shorn lamb. In case after case the amercement is pardoned because he or she is poor, because he has nothing.[4] There is a humane side to the picture which is too often overlooked.

[1] f. 6; Vinogradoff, op. cit., p. 74 f.
[2] *C.R.R.* i. 382.
[3] Cf. *The Earliest Northamptonshire Assize Rolls*, nos. 859, 917, 931; *C.R.R.* ii. 295.
[4] See the examples collected by Mrs. Stenton in the introduction to the *Rolls of the Justices in Eyre for Lincolnshire 1218–19 and Worcestershire 1221* (Selden Soc., vol. liii), pp. lxiii–lxvi. She concludes: 'At Worcester and Lincoln, doubtless remembering the limitations upon amercement prescribed by the great charter, they (the judges) are obviously careful not to impose a burden beyond anyone's financial capacity' (p. lxvi).

VI

FEUDAL INCIDENTS

I N the Middle Ages the words *pax* and *finis*, like the word *misericordia* which I have already discussed, have sometimes a sinister implication. They seem innocent enough if we translate them as 'peace' and 'end', but more ominous and alarming if we translate them, as we fairly can, as 'payment' and 'fine'. The word *finis* in the Exchequer always implies making an end of a matter by a composition in money or kind; the word *pax* is used in both senses. Thus the king writes to his justiciar Geoffrey Fitz Peter: 'Know that of the 200 marks for which Robert of Tresgoz fined (*finivit*) with us for the lands of William, he has paid us (*pacavit*) 50 marks.'[1] A good illustration of the double use of the word *pax* is afforded by a letter of the same Geoffrey Fitz Peter to the sheriff of Essex: 'If Gerard of Maldon shall have given you security of making payment (*faciendi pacem*) to your account at the Exchequer of 100s. which you require from him, then we order you to leave him in peace (*ei pacem habere facias*) until that term.'[2] Behind these words *pax* and *finis* lies a threat; often it amounts to blackmail on the part of the Crown against its subjects. In effect the king says: 'I will destroy you, seize your land and personal property unless you satisfy me with a payment and make an end of the matter.' A man gets himself into trouble, and makes an end of the quarrel or dispute by an offer (*oblatum*) of a sum of money; he *facit finem cum rege* for so many pounds, so many marks; he promises to make payment (*affidavit facere pacem*) at the Exchequer of so much money for having the king's goodwill.[3] These offers (*oblata*) are recorded on a special roll—*Rotulus de Oblatis et Finibus*[4]—and then accounted for on the great roll of the Exchequer.

The Angevin kings were notoriously quick-tempered; the

---

[1] *Liberate Roll 2 Jo.* (Pipe Roll Soc., N.S. xxi); p. 90; cf. *Book of Fees*, pp. 538–9. *Plene pacavit ut dicit* against entries of scutage dues.

[2] *Memoranda Roll 1 Jo.*, p. 83.

[3] Ibid., p. 78.

[4] The surviving rolls for the reign of John were edited by T. Duffus Hardy for the Record Commission in 1835.

most faithful subjects might for little fault of their own earn the king's displeasure. But his goodwill could generally be restored by a suitable present in money or kind. Every year substantial sums were paid into the treasury 'for having the king's benevolence', 'for having the king's love', 'for having the king's peace', or that 'the king's anger might be relaxed'. For a price the king would interfere in the course of justice, would sell his mediation between two parties, or would meddle in the most intimate domestic concerns of his people. King John, whose greed has not been exaggerated by historians, made the strangest bargains. Andrew Neulan, a Londoner, offers him three rain cloaks (*capas pluviales*) from Flanders for his intercession with the prior of Chicksand to keep an engagement;[1] Roger son of Nicholas offers as many lampreys as he can get hold of, if the king will induce William Marshal to grant him the manor of Langford;[2] William de Braose gives ten bulls and ten cows so that he need not go up to Scotland to fetch the king of Scotland to court;[3] William de Vallibus, who has apparently committed an indiscretion with the wife of Henry Panel, offers the king five of the best palfreys that he, the king, will say nothing about it.[4] Perhaps the most curious entry on the Fine Rolls of the reign of King John is one that records that 'the wife of Hugh de Nevill gives the lord king 200 hens that she may lie one night with her husband', and Hugh de Nevill himself and another responsible person go surety that the birds will be safely delivered before the next Easter.[5] King John, who, whatever his failings, did not lack a sense of humour, was obviously pleased with this render of 200 hens, for the next entry but one on the Fine Roll records a similar offer from Stephen of Oxford for a letter from the king praying Alice, the widow of John Kepeharm, to take him as her husband.[6] The lady appears to have assented, for on the Pipe Roll of the same year she renders account of 100 marks and two palfreys that she may marry according to the laws and customs of the city of Oxford (what these special Oxford marriage customs were

---

[1] *Rot. de Fin.*, ed. Hardy, p. 414.
[2] Ibid., p. 511.
[3] Ibid., p. 334.
[4] Quoted Hardy, ibid., p. v, n. 1. The entry appears to have been omitted from the text.
[5] Ibid., p. 275.   [6] Ibid.

does not appear) and to have what she ought to have of her late husband's lands and chattels.[1]

Such instances show the diverse nature of fines. The sums of money exacted by the king by way of relief when a tenant entered into his estate or in connexion with wardship and marriage are also of the nature of fines; they conclude a bargain; they are compositions of the king's claim to take what he pleases.

These obligations incidental to feudal tenure, the feudal incidents as they are called, reflect a time when the king must control the private as well as the public lives of his subjects. These men were endowed with wealth and power which the king could not allow to be transmitted to untrustworthy hands. This was one of the underlying motives governing the system of reliefs, wardship, and marriage. It could, however, in the hands of arbitrary and unscrupulous kings be easily turned into a system of blackmail. The prominence given to these matters in the charters of liberties and in Magna Carta, the last and greatest charter of liberties, shows how deeply the barons felt about them and how much they were abused.

An heir could not enter upon his inheritance without first making a bargain with his lord; he must pay a relief, a succession duty. The barons insisted that this relief should be 'just and lawful', and this Henry I promised in his coronation charter. But what was a just and lawful relief? A compilation known as the *Leis Willelme*, the so-called laws of William the Conqueror, written according to Liebermann not later than 1135 and possibly as early as 1090, gives us a hint.[2] The author strangely confuses the Anglo-Saxon heriot and the Anglo-Norman relief. Most of the relevant chapter is taken from Cnut's law of heriots which has to do with horses, arms, and armour. Speaking of the relief of a vassal to his liege lord, he says it is his father's horse, breastplate, helmet, shield, lance, and sword; but, he continues, if he has not got these things, he can acquit himself by the payment of 100s. The relief of a man who holds his land by an annual rent shall be a year's rent. We are probably justified in assuming, therefore, that

---

[1] *P.R. 7 Jo.*, p. 151.

[2] c. 20. *Die Gesetze der Angel-Sachsen*, ed. Liebermann, i. 507. For the date of the collection, ibid., p. 492 note.

already in the time. of Henry I the 'just and lawful' relief of a tenant by knight service was understood to be 100*s*., and that of a tenant in socage a year's rent. The surviving Pipe Roll of this reign, that of the year 1130, does not afford sufficient evidence to draw any conclusion about the practice in the matter of reliefs beyond the fact that they seem to be in excess of what was 'just and lawful'.[1] In the time of Henry II the question is definitely settled. Both Richard Fitz Nigel in the *Dialogue of the Exchequer* written about 1179,[2] and Glanvill, less than ten years later,[3] regard it as established. The former says 'the heir being of age pays one hundred shillings for each knight's fee, or less, that is 50*s*., if he possesses half a knight's fee, and so on'. It is evident from the Pipe Rolls that Henry II kept to this rule. Thus, to take a few examples, Miles of Beauchamp paid 50*s*. on half a fee in Buckinghamshire; John Puherus £5 for one fee in Northamptonshire; Hugh Burdet £10 for two fees in Leicestershire; Brian Fitz Ralph £25 for five fees in Suffolk. These are all taken from the fiscal year 1185,[4] and prove conclusively that reliefs of tenants holding of the Crown by knight service were charged at the rate of £5 per fee.

The king, however, had his chance when dealing with barons. Speaking of these the author of the *Dialogue* says 'he is not to satisfy the king according to a settled sum, but as he may be able to arrange with the king'. The precise technical meaning attached to the word *baronia* in this period is still far from clear, but it is certain that if a tenant held his estates *per baroniam*, the king could exact an arbitrary relief. Though the 1,000 marks charged against Robert de Lascy in 1178 was perhaps unusually high,[5] a relief of £200 was very common. Walter Brito was charged this sum for a barony of fifteen fees and William Bertram the same amount for one of only three.[6] Had these tenants held of the Crown by knight service and not by barony they would have only had to pay £75 and £15 respectively.

---

[1] On the other hand, Professor Stenton (*English Feudalism*, pp. 162–3) cites evidence to show that the mesne tenants were paying reliefs at a considerably lower rate than 100*s*.

[2] ii. 10.   [3] ix. 4.   [4] *P.R. 31 Hen. II*, pp. 140, 53, 104, 41.

[5] *P.R. 24 Hen. II*, p. 72. The Pontefract fief of the Lascy family was a large one; it was returned in 1166 as sixty knight's fees (*Red Book of the Exchequer*, p. 421).

[6] *P.R. 11 Hen. II*, p. 65; *P.R. 23 Hen. II*, p. 83. *Red Book*, pp. 232, 442.

There was, however, a general understanding that the relief
of barons should be 'reasonable' (*rationabile*). This is insisted
upon by Glanvill.[1] The term 'reasonable' is certainly very
vague. But in fact it was well understood to mean a definite
sum. On the Pipe Roll of 1198 it is stated that 'William de
Neufmarché renders account of 100 marks that the king may
take his reasonable relief, *scilicet* £100'. The next entry reads:
'The same William renders account of £100 for the relief for
the land of his father.'[2] By a 'reasonable' relief then was understood £100; but in order to get off with a reasonable relief,
this tenant had to offer the king a bribe of an additional 100
marks. This entry on the roll of 1198 explains the wording of
the second chapter of Magna Carta, which reads 'he shall have
his inheritance by the ancient relief; *scilicet* the heir or heirs of
a barony £100 for the whole barony'. It has been too readily
assumed that extortion in the matter of reliefs was introduced
or at least heavily increased by King John. The practice was,
as we have seen, one of long standing. There are certainly
instances of great and ruinous extortion in his reign; he required
the vast sum of 10,000 marks from Nicholas de Stutevill for the
inheritance of his brother in 1205.[3] But this is exceptional; on
the whole he was no more excessive in his demands than his
predecessors. He would often content himself with the reasonable relief of £100 for a barony.

The relief was perhaps less subject to abuse than other royal
rights of this kind. Of all the burdens attached to feudal
tenures, none were so irksome as those connected with the rights
of wardship and marriage. These applied to tenants holding
by military service or by sergeanty. The lord enjoyed the profits
of an estate during the minority of the heir on the ground that
the heir by reason of his youth was incapable of rendering any
service. This extended in the case of boys till the age of twenty-
one and in that of girls till the presumed marriageable age of
fourteen. The lord also controlled the marriages of his wards.
Although Glanvill[4] speaks only of the marriage of female
wards, the records of the late twelfth century supply abundant
evidence that the lord exercised the right over male wards as

---

[1] ix. 4.   [2] *P.R. 10 Ric. I*, p. 222.
[3] *P.R. 7 Jo.*, p. 59. The money in fact was never paid. See Farrer, *Early Yorkshire Charters*, i. 391 f.   [4] vii. 12.

well. He disposed also of widows. In the matter of marriage the aristocracy was far more seriously restricted than the peasantry. Indeed, it is somewhat curious that the payment of a small sum, a few shillings, by the villein for leave to give his daughter in marriage—the merchet—is always regarded as a peculiar mark of his degraded position, the supreme test of his servility, whereas the payment of a hundred or a thousand pounds by a baron's widow that she may marry whom she chooses is regarded as quite normal and what must be expected from persons of exalted rank. But in essence there is little difference between them. Both illustrate the dominating influence of the lord in feudal society.

Abuses of the system of wardship and marriage were of long standing. Henry I promised reform; he would take no money for the licence to marry, nor would he refuse it, unless the proposed marriage was with one of his enemies. He would entrust wards to the guardianship of a near relation.[1] But like most else in this first charter of liberties, these concessions went for nothing. Widows and orphans continued to be sold for large sums. The largest recorded in this period was the sum of 20,000 marks owed by Geoffrey de Mandeville 'for having to wife Isabel countess of Gloucester with all her lands, tenements, and fees which belong to the same Isabel, except the castle of Bristol and the chases outside Bristol'.[2] This, however, was exceptional. We should expect her to be expensive. She was the divorced wife of the king, and she brought to her husband an earldom and considerable wealth. Nevertheless, it was a great sum. According to the estimate of Sir James Ramsay, the whole ordinary revenue of the kingdom for this year (1214) was not much above £42,000.[3] But prices in the early years of the thirteenth century were high and rising. It was not unusual to give a sum of four figures for a widow. Gerard de Camville, for example, gave £1,000 for the widow of Thomas of Verdun for his son.[4]

These fines must have been a serious drain on the estates of heiresses. But a greater grievance was one not of money but

[1] Coronation Charter, cc. 3, 4.
[2] *Rot. de Fin.*, p. 520. No part of this debt incurred in 1214 was paid by 1218; see Madox, *Exchequer* (ed. 1711), p. 322 note *uu*.
[3] *Revenues of the Kings of England*, i. 261.
[4] *P.R. 2 Jo.*, p. 87.

of class. What these aristocratic ladies most feared was that they would be forced to marry beneath them. They might be given by the king to his officials, men of humble origin who had risen to lucrative and responsible positions in the king's service. They might be disparaged, that is to say, married not with their *pares*, their equals.

Disobedience of the king's orders in respect of marriage was severely punished. Sometimes a threat is introduced into the agreement: 'if that widow should refuse him' he (the proposed husband) shall have all her land.[1] Nor were these empty threats. In 1204 Ralph Ridel gave 50 marks and two palfreys for Alice the widow of John Belet. He was apparently doubtful of his success in getting the lady, for he contracted to have his money back if the marriage did not take place.[2] The next year we hear that the king has taken her dower lands into his own hands because she refused to marry according to his will.[3] The matter was ultimately settled by her father paying 100 marks for having the marriage of his daughter and that she might have her dower.[4] Three years later she married Thomas de Burgh.[5]

Women would pay substantially for their independence in the matter of marriage and that they might not be disparaged. In 1205, for instance, Mabel, the widow of Hugh Bardolf, offered 2,000 marks and five palfreys 'that she be not constrained to marry herself, and that she may remain a widow as long as she pleases' and for certain other privileges.[6] The countess of Warwick in the same year offered £1,000 and ten palfreys 'that she may remain a widow and not be forced to marry' and for having the custody of her sons and their lands.[7] Many others paid smaller sums 'that they may marry whom they please' or 'that they be not compelled to marry against their will'. The eighth chapter of Magna Carta was obviously intended to remedy this particular abuse. 'No widow', it declares, 'shall be compelled to marry so long as she wishes to live without a husband.' This went some way towards removing the grievance; but nothing is said about payment for the privilege of remaining single. In fact payment was still demanded, but at a much more reasonable figure. Whereas in 1205, as we

---

[1] *Rot. de Fin.*, p. 40. [2] Ibid., p. 226. [3] *C.R.R.* iii. 257.
[4] *Rot. de Fin.*, p. 287; *P.R. 7 Jo.*, p. 211.
[5] *Rot. de Fin.*, p. 440. [6] *P.R. 7 Jo.*, p. 34. [7] Ibid., p. 33.

have seen, 1,000 marks and ten palfreys were demanded from the countess of Warwick for the privilege of remaining a widow, no more than 100 marks were required from her successor in title, Philippa, the widow of Henry, the fifth earl, 'that she be not constrained to marry so long as she prefers to live without a husband, and that she can marry whom she pleases'.[1]

Even if a woman bought her freedom, she was not altogether out of the clutches of the Crown agents. William of Clinton, lord of the manor of Cassington near Oxford, died in 1197. His widow Isabel quickly purchased her right to marry as she pleased.[2] The little estate—it was only a single knight's fee— was farmed for the Crown by one Walter Fitz Godfrey who, it seems, farmed it honestly. It was worth £5; and in the first year he spent 2 marks on the repair of Cassington mill, he gave £3 to Isabel for the maintenance of her sons, and paid the odd mark into the treasury.[3] So it went on the next year, £3 to the boys and the balance to the treasury.[4] Then at Easter 1200 the wardship with the marriage of the heir was sold to a Buckinghamshire knight, Hugh of Haversham, for 200 marks.[5] The widow, thus threatened, was forced to bid higher; and she offered 300 marks and a palfrey 'for having the custody of the land and heir and the right to marry him at her will so that he be not disparaged'.[6] Her offer was accepted, and the sale to Hugh of Haversham was cancelled.[7]

Sales in this expensive marriage market were not always done by private treaty. Sometimes there was brisk bidding. Richard de Lec offered 80 marks for a widow; Simon of Kyme gave £100.[8] Hugh de Nevill the forester offered 30 marks for the custody of the land and heir of Godfrey of Standen; Godfrey of St. Martin offered 50, and won.[9] Perhaps the most unscrupulous example of sale by auction comes from a memorandum on the Fine Roll of 1201:[10] 'Hugh de Nevill offers the king 30 marks for a certain marriage for the use of his grand-daughter, and if anyone is prepared to give more for that wardship than the said Hugh de Nevill, let him have it, unless Hugh de Nevill is willing to give as much.' But in

---

[1] P.R. 14 Hen. III (1230), p. 214.　　[2] P.R. 10 Ric. I, p. 195.
[3] Ibid., p. 194.　　[4] P.R. 1 Jo., p. 224.
[5] P.R. 2 Jo., pp. 23, 265.　　[6] Ibid., p. 27.
[7] Rot. de Fin., p. 61.　　[8] Ibid., pp. 51, 56.
[9] Ibid., pp. 174, 175; P.R. 3 Jo., p. 81.　　[10] Rot. de Fin., p. 192.

fairness to the king, it must be allowed that he did not always accept the highest bid. William Bardolf in 1196 promised 200 marks to marry the daughter of Almaric Dispenser.[1] In 1198 Peter of Stokes offered 100 marks.[2] On the Pipe Roll of the next year, after reciting William Bardolf's offer, the words are added: 'But it is recorded by Geoffrey Fitz Peter that King Richard was unwilling to accept that fine, and that he has given the said daughter of Almaric to Peter of Stokes by the fine of 100 marks which Peter made with him.'[3]

We can derive a fairly clear picture of the system of wardship and marriage at work from a unique roll of the year 1185 known as 'the Ladies Roll', or, to give it its full title, 'Rolls of ladies, boys, and girls in the king's gift' (*Rotuli de dominabus et pueris et puellis de donatione regis*).[4] When itinerant justices visited the shires they were furnished with a list of subjects on which the Crown required information, the articles of the eyre as they were called. In the earliest of these lists that has been preserved, that of the year 1194, were included inquiries 'concerning wardships of boys which belong to the king' and 'concerning marriages of girls and widows which belong to the king'.[5] The 'Ladies Roll' supplies the answers from twelve counties to similar inquiries made in 1185. Here is a typical answer taken from the returns relating to Northamptonshire:

'Alice, who was the wife of Fulk de Lisures and the sister of William de Auberville, is in the gift of the lord King, and she is fifty years of age; and she has two sons knighted and two others, and six married daughters and three marriageable daughters who are in the guardianship of the mother. Her land in Glapthorn is worth 100*s.* with this head of stock, viz. two plough-teams and six cows and one bull, and 30 pigs and 40 sheep. Her land in Abington, which is her dowry and is in the hundred of Spelhoe, is worth £14 a year.'[6]

This information was required for two reasons. First, that the Crown might know what price could fairly be placed upon the widows and children in the marriage market; and secondly the king wished to be assured that the estates in his custody

---

[1] *Chancellor's Roll 8 Ric. I*, p. 39.
[2] *P.R. 10 Ric. I*, p. 107.
[3] *P.R. 1 Jo.*, p. 7.
[4] It is printed by the Pipe Roll Society, vol. xxxv.
[5] Roger of Hoveden, *Chronica*, iii. 263 (articles 5 and 6).
[6] *Rot. de Dom.*, p. 24.

were being farmed in the most profitable manner and being kept properly stocked.

For the first of these purposes it was necessary to know the age of a widow, the number of her children, and the value of her dowry. Sybil of Harlton, we are informed, is in the king's gift. She is the daughter of Roger of Gigney, and is upwards of seventy years old; besides the heir, she has a string of nine other young children (*infantes*); and her whole property is worth but £10 a year.[1] Obviously with her the king has not got a very saleable article. The husband would not be likely to enjoy her dowry for long, and the young children might have to be provided with marriage portions (*maritagia*). On the other hand, Matilda, the daughter of Thomas Fitz Bernard, whose first husband, John of Bidun, left her a widow at the age of ten,[2] could be readily disposed of at a good price. She might reasonably expect to enjoy her dowry at Stow and Kirby Bedon in Norfolk for a considerable time. She lived, in fact, with her second husband, John of Rochford, till 1255, seventy years after her first husband's death. It is not always easy even with these precise records to work out the family history. The younger ladies willingly gave their exact ages; one is eighteen, another twenty-four, and one little heiress only five (who at this tender age was given in marriage to the son of Thomas Fitz Bernard). But widows of middle age and upwards were reticent in the matter. The justices in making their returns have often to hazard a guess; they enter round numbers; she is fifty, sixty, seventy; sometimes they add 'and a bit more' (*et amplius eo*). Even a middle-aged widow might, however, prove a valuable match. Of Mary, the widow of Guy L'Estrange, we read:[3] 'She is forty years old and born of knights and barons. She had three husbands. Her dower lands and marriage portion are in diverse counties.' She may have acquired a dowry from each of her three husbands.

If we now turn to the other side of the inquiry made by the justices, the farming of the estates in wardship, we see a picture of appalling neglect, mismanagement, and shameless profiteering by the Crown agents to the injury of the heirs under their

---

[1] Ibid., p. 85.
[2] Ibid., pp. 49, 55; cf. introd., p. xxxvii, and Farrer, *Honors and Knights' Fees*, i. 2 f.
[3] *Rot. de Dom.*, p. 53.

charge. These unjust stewards made what they could out of the estates over and above the legal profits, the farm. In 1178 Gilbert de Monte withdrew from the world into the abbey of Eynsham. For the next eight years his son and heir, his four daughters, and his estate at Whitfield in Northamptonshire were in the custody of the sheriff, Thomas Fitz Bernard.

'The land' (the justices report) 'is worth annually £4. 16s. 5d. of assized rent, and this rent Thomas Fitz Bernard received while he had the custody. But if it were reasonably stocked, that is to say, with two plough-teams, 100 sheep, four cows and a bull, and four sows and a boar, it would be worth £8. 10s. 5d. But there are there now only the two plough-teams of the said stock. Further, John Clerk who then was steward of Thomas Fitz Bernard gave to the township of Whitfield forty yearling sheep at Martinmas, and received in exchange as many good ewes with lambs at Easter, unjustly. And besides he took 16 casks of beer, worth twopence apiece, for which he gave nothing; moreover he took one ox, and Ralph Morin half a mark, from a man of the village. The agents of Thomas Fitz Bernard took during these eight years £5. 18s. 0d. in addition to the farm, unjustly.'[1]

This was a bad case. It is unusual for the justices to include in their official report so outspoken a rebuke as is implied in the word *injuste* with reference to these transactions.

But this does not stand alone. An even more disgraceful example of neglect is revealed on the Vesci manor of Caythorpe in Lincolnshire. William de Vesci died in 1182, and the manor was entrusted to one Adam Fitz Robert, who mismanaged it so seriously that he had speedily to be removed. The justices tell us the state of the property when Adam of Carlisle, the next custodian, took over:

'He found there 28 oxen of which one was dead; and five draught horses of which one was dead; seven yearling pigs and eight smaller ones; and 600 sheep of which 260 had been sold at £10. 11s. 8d.; of the remainder 280 were dead and their fleeces were sold for £4. 12s. 0d.'[2]

As these sums are not entered on the account which Adam of Carlisle rendered to the Exchequer,[3] it is probable that they found their way into the steward's pocket. Adam in his turn

---

[1] *Rot. de Dom.*, p. 29.　　　　　　　　　　　　[2] Ibid., pp. 9 f.
[3] *P.R. 30 Hen. II*, pp. 154 f.

gave way to another custodian, Hugh of Morewich, who took over a still more diminished head of stock; another ox and another horse have died; and the flock of sheep is reduced to fifty. Hugh of Morewich was one of the justices who conducted the inquiry of which we have the record in the *Rotuli de Dominabus*. He was probably a good manager, for we find on the Pipe Roll of this year 1185 a very large sum expended on the repair and the stocking of the manors of the late William de Vesci.[1] Another and a more irreparable form of injury may be illustrated from the return relating to the rich manor of Kimbolton in Huntingdonshire. This manor of William de Say had been in the hands of the Crown for seven years at the time of our record, and was in the custody of Richard Ruffus. Besides helping himself on a lavish scale to the farm stock (he is said to have removed 20 oxen, 5 cows, and 3 horses) he took from the wood on the estate 222 oak-trees, 42 of which he distributed among his friends, while the remainder he used to build himself 'a hall and a chamber' in Leicestershire.[2]

I have dwelt at some length on the administration of wardships as revealed by the *Rotuli de Dominabus* because it illustrates very vividly the grievances of which the barons complained and which they had redressed in the fourth and fifth chapters of the Great Charter. These articles were directed against destruction and waste: the guardian of an estate must take nothing but the reasonable produce, and that without destruction or waste; and conversely, he must hand it back to the heir at the end of his stewardship in good repair, properly stocked and cultivated. It is manifest from the returns to the inquiry of the justices in 1185 that the grievances complained of were not only very genuine but of long standing. King John was merely carrying on a practice already much abused by his father. This evidence enables us to understand why the barons of 1215 gave such emphasis to the feudal incidents of reliefs, wardship, and marriage. They are given the most prominent position. The first four 'Articles of the Barons' relate to them; they are the subject of chapters 2 to 8 of the Great Charter itself. They were at the forefront and fundamental in the baronial programme of reform.

The kings of this period would exact money from their

---

[1] *P.R. 31 Hen. II*, pp. 8 f.   [2] *Rot. de Dom.*, p. 46.

subjects on the slightest pretence in the shape of fines and
amercements; they would demand what they could for reliefs,
wardships, and marriages. But there is a more pleasant aspect
of the business; the appearance is more alarming than the
reality. The terms of payment were usually light. To-day, if
a man gets into trouble in the courts, he will get short shrift;
he must pay quickly or go to prison. This was not so in the
twelfth and thirteenth centuries. Whether it was because the
machinery of collection was inadequate or because the officials
were intentionally lenient, the fact remains that men were
allowed to pay what they owed in very easy instalments. In
1185, for instance, Roger, an Oxford cook, incurred a fine of
40 marks 'for hanging a man without the view of the justices
and sheriff'.[1] For the first seven years he paid nothing at all;
indeed, 10 of the 40 marks were pardoned in 1187.[2] In 1192
he paid a mark, and the next year half a mark. For the next
three years he paid 4 marks, and in 1197, 2 marks; each subsequent year until 1201 he paid a mark. Then in 1202 Roger
apparently died. The unpaid balance now stood at 10 marks,
and against this sum there is a note 'for which Otto son of
Roger should render ½ mark a year'.[3] This he accordingly did,
and at last, presumably in 1222, thirty-eight years after the
fine had been imposed, it was finally paid off. An even more
striking instance is afforded by the case of Geoffrey the Bursar
who got himself into trouble in 1165 and incurred a fine of
200 marks.[4] He had paid nothing when he died in 1169.[5] The
account was then transferred to his son Alan, who from time
to time paid a mark or so into the treasury. In 1191 it was
arranged that he should render a mark each year;[6] and this
he was still doing in 1230. The balance then stood at 152
marks.[7] At this rate of progress the original fine of 200 marks
would be finally settled in 1382, the sixth year of the reign of
Richard II, 217 years after the debt had been incurred.

This system of payment by easy instalments is apt to mislead
the unwary. Sometimes the fact that it is a balance and not
the original debt which is entered on an account is indicated

[1] P.R. 31 Hen. II, p. 108.
[2] P.R. 33 Hen. II, p. 47.
[3] P.R. 4 Jo., p. 205.
[4] P.R. 11 Hen. II, p. 32.
[5] P.R. 15 Hen. II, p. 171.
[6] P.R. 3 Ric. I, p. 137.
[7] P.R. 14 Hen. III, p. 98.

by phrases such as 'as is noted on the preceding roll' or 'as is contained in the roll of the third year'. But more commonly there is no such indication. For this reason it is most unsafe to use Pipe Rolls in isolation. Even Dr. Round, who edited many Pipe Rolls, sometimes fell into this elementary trap. This is so strange in one who dealt so savagely with careless editors that I am tempted to quote a somewhat startling instance. Commenting in his introduction to the roll of 1178 on the arbitrary character of the baronial relief he says: 'William de Montacute pays 100 marks for his father's fief, which was one of ten knights' fees; on the same page Walter Brito is found charged but forty marks for his relief, though his "service" was that of fifteen knights.'[1] In fact Walter Brito had succeeded to his estate thirteen years before, and if we turn to the roll of 1165 we find him charged with a relief not of 40 but of 300 marks.[2] He has been slowly paying it off, and by 1178 there is still an unpaid balance of 40 marks owing. It was this which Round mistook for the original relief.[3]

It was to avoid debts running on almost interminably that towards the close of the twelfth century it became customary, at least when large sums were involved, to make conditions about payment. I have already referred to the case of Alice, countess of Warwick, who proffered £1,000 and ten palfreys that she might remain a widow and have the custody of her land and children. After reciting the privileges which she was to enjoy, the entry on the roll specifies in precise terms the conditions of payment:[4]

'On the Purification of the Blessed Mary in the sixth year she shall render two palfreys not in money but in palfreys; and at mid-Lent £100 and three palfreys not in money; at the Easter Exchequer 100 marks and two palfreys in cash; on the feast of St. John 50 marks and one palfrey in cash; on the feast of St. Michael £100 and two palfreys in cash; and so from Exchequer to Exchequer at each £100 until the said debt is paid.'

[1] *P.R. 24 Hen. II*, p. xxii.
[2] *P.R. 11 Hen. II*, p. 65.
[3] In the same introduction he speaks of Everard de Ros paying off the 800 marks claimed from him on succeeding to his father. This 800 marks was in fact the balance of a debt of 1,000 marks incurred when his father Robert de Ros succeeded to the estates of Walter Espec (*P.R. 4 Hen. II,* p. 146). The son took over the liability in 1163 (*P.R. 9 Hen. II*, p. 58).
[4] *P.R. 7 Jo.*, p. 33.

There follows a note of the first payment:

'In the treasury £100 and five marks for one palfrey. And to the king himself £100 in his chamber; and five palfreys; and she owes £800 and four palfreys of which she must render £200 a year.'

The king also took a further precaution. The countess must find sureties for the payment. On the Fine Roll there is a list of 34 persons who, in sums varying from £200 to £1, are prepared to guarantee the punctual payment of the debt.[1]

This was usually the extent of the pressure put upon Crown debtors. The barons of the Exchequer would generally take a lenient view when serious cases were brought up for consideration by the Remembrancers. They would readily grant a postponement to a given term; they would accept almost any excuse. Richard de Herierd, whose debt was already three years overdue, pleaded for a respite 'because he says he is always prompt'.[2] The king could, like a private person, distrain his debtor by the seizure of his land or chattels. But he seems to have seldom had recourse to such extreme measures;[3] and when he did, it was with little effect. Andrew Talebot owed the king £10 in 1196. Three years later, as nothing had been paid, a writ was issued to distrain him.[4] It was probably not put into execution, for the amount continues to be entered against him till 1205, when he begins slowly to pay it off.[5] There was no imprisonment for debt in the twelfth century. So debts were allowed to run on year after year. Sometimes the officials at the Exchequer get tired of entering these unpaid debts, and they are pardoned or silently dropped.

The full significance of this system of payment by small instalments will be realized if we compare the sums due to the Exchequer in any year with what was actually received. In the Exchequer year ending at Michaelmas 1200 from the counties of Norfolk and Suffolk there was due from scutages, aids, tallages, fines, and amercements, from every source, that is to say, except the county farm and escheated honours, the sum of £7,215. 18s. 11d. Of this sum £1,132. 4s. 8d., or

---

[1] *Rot. de Fin.*, pp. 276 f.   [2] *Memoranda Roll 1 Jo.*, p. 19.

[3] By Henry III's time distraint for debt was in common use. Entry after entry on the Memoranda Roll of 1230 (Pipe Roll Society, N.S. vol. xi) is followed by a note *distringatur* or simply *dist.*

[4] *Memoranda Roll 1 Jo.*, p. 66.   [5] *P.R. 7 Jo.*, p. 134.

approximately 16 per cent., was actually received by the treasury officials.[1] The figures for Oxfordshire are still more striking. Of £1,191. 0s. 7d. entered on the account of the same year, no more than £142. 10s. or about 12 per cent. was in fact paid into the treasury.[2]

The Anglo-Norman financial system was unrivalled in Europe. But it may be questioned whether, like the equally unrivalled judicial system, it worked so effectively in practice as it would seem to do in theory. There is either great leniency or great slovenliness in the work of the officials. Revenue officers to-day would not remain long at their posts if they could collect no more than 12 per cent. of the taxes for which they were responsible. The kings of this period were very extortionate; the pecuniary burdens seem to be among the heaviest that all classes of society had to bear. But these burdens were very much mitigated by an almost universal system of pay as you please without the fear of a debtors' prison in the background.

[1] *P.R. 2 Jo.*, pp. 129–49. In this roll a few entries at the end are deficient. No account has been taken of the value of 5,000 eels which made up part of the proffer of the burgesses of Dunwich (p. 147) or of various sparrow hawks. Palfreys have been reckoned at the current rate of 5 marks. Two payments were rendered not into the Exchequer but into the king's *camera* (p. 146).

[2] Ibid., pp. 21–8.

INDEX

Abbots Bromley (Staffs.), 25 f.
Abingdon, abbey of, 49.
Abington (Northants.), 100.
Adewich, Ralph son of William of, 6.
Ailnoth the engineer, 64 f.
Ailric, Roger son of, 85.
Ailwin, priest, 29.
— Peter son of, 18.
Alexander the carpenter, 70 n. 5.
Alfred, King, 2.
Allerthorpe (Yorks.), 30.
Almaric, Count of Evreux, 33, 34 n. 1.
Alneto, Henry de, of Cornwall, 53 n. 4.
— — of Maidford, 53 n. 4.
Aluric, priest, 29.
Alvescot (Oxon.), 74.
amercements, 77–91; roll of, 82, 84 f.
ancient demesne, 17 f., 26.
Aneford (Andoversford, Glouc.), Ricardus de, 9.
Anne, Queen, 49.
Appleby (Leics.), 25.
Apps in Walton-on-Thames (Surrey), 57.
aquarius, 78 and n. 2.
Arngar, 30.
Articles of the Barons, 103.
Assize of Arms, 33.
Aston, Middle (Oxon.), 74.
— Rowant (Oxon.), 61, 74.
Auberville, William de, Alice sister of, 100.

Bagshot (Berks.), 49.
Baldwin, William son of, 31.
Bampton (Oxon.), 68.
Bardolf, Hugh, Mabel widow of, 98.
— William, 100.
Barksdon Green (Herts.), 29.
Barnstaple, 35 n. 2.
Basset, Ralph, 88.
Baus, William de, 55 n. 2.
Bayeux, Hugh of, 43.
Beauchamp, Miles of, 95.
— William of, 5.
Beaumont, John son of Philip of, 19.
Becco, Walter de, 36.
Belet, John, Alice widow of, 98.
Bericote (War.), 62.
Berkshire, 49.
Bertram, William, 95.
Bessacar (Yorks.), 6.
Bidun, John of, 101.
Bifeld (Byfield, Northants.), 8 n. 1.

Billingshurst, William of, 48.
Bladon (Oxon.), 68.
Blechingdon (Oxon.), 69.
Blundus, Robert, 46.
Bodiham, William de, 35 n. 2
Boilliers, Robert de, Philip knight of, 51.
Bolebec, Herbert de, 60.
Book of Fees, 1, 38.
Boscher, 62.
Bosco, Robert de, 76 n. 4.
Bracton, 2, 3, 16, 17, 18, 19, 20 f., 25, 28, 55 n. 6, 56 and n. 4, 59, 84, 91.
Braishfield (Hants), 40.
Brampton (Suffolk), 47.
Brakelond, Jocelin of, 41, 48 f.
Braose, William de, 93.
Bremûle, battle of, 37.
Bristol, 51, 97.
Brito, Walter, 95, 105.
Broc, Nigel of, 40.
Broughton Poggs (Oxon.), 68.
Buckinghamshire, 60.
Budde, Anderd, 27.
Buffin, family of, 71.
— Thomas, 38, 71 n. 8.
— William, 71 and n. 8.
Bugedenn, William of, 9.
Buis, Nicholas le, 19.
— Warin, son of Guy le, 19.
Burdet, Hugh, 95.
Burgh, Thomas de, 98.
Burton, abbey of, 24, 25.
— Laurence, abbot of, 26.
Bury St. Edmunds, 41, 48 f.
Bussel, Robert, 58.
Bustardthorpe (Yorks.), 62.
buzones, 56 and n. 6.

Caen, abbey of the Holy Trinity at, 24.
Cam, Miss H. M., 22 n. 2.
Camville, Gerard de, 58, 97.
Canterbury, 39.
— abbot of St. Augustine's, 43.
— archbishop of, 64.
Canterel, Robert, 34.
Carlisle, Adam of, 102.
Cassington (Oxon.), manor of, 99; mill, 99.
castle guard, 37, 38, 48 f.
Caythorpe (Lincs.), 102.
Cedric, 18; Saiva, wife of, 18.
censarii, 25.
Chancels, Guy de, 8.
Charles II, King, 45.
Chauncy, Simon of, 5.

PRINTED IN GREAT BRITAIN
AT THE UNIVERSITY PRESS, OXFORD
BY VIVIAN RIDLER
PRINTER TO THE UNIVERSITY

Date Due

OCT 30			
SEP 26 1973			
DEC 15 1995			
NOV 10 1996			
NOV 0 3 1996			

18998

CPSIA information can be obtained
at www.ICGtesting.com
Printed in the USA
BVHW050540140223
658474BV00020B/255